Cutter Confusion

Cutter Confusion

Trevor Calcott Walker

Illustrated by the Author

The Pentland Press Limited
Edinburgh • Cambridge • Durham

© Trevor Calcott Walker 1993

First published in 1993 by
The Pentland Press Ltd.
1 Hutton Close
South Church
Bishop Auckland
Durham

All rights reserved.
Unauthorised duplication
contravenes existing laws.

ISBN 1 85821 117 4

Typeset by CBS, Felixstowe, Suffolk
Printed and bound by Antony Rowe Ltd., Chippenham

To

Alice and Harry

LIST OF ILLUSTRATIONS

B.O. Davies	Frontispiece
Stand by the anchor	17
A bunch of characters on the boatdeck	46
I was suddenly conscious of flames behind me	52
Towing was better than rowing	57
We craned our necks to catch the heavy falls wildly thrashing about above our heads	59
We're sinkin'! We' sinkin'!	61
The relief comes aboard	64
The Swedish turret ship *Vindo*	66
Blimey! He's kicked the bucket!	105
No. 4 Watch	106
Seeking Greeks	113
Face each other on the bare boards of the hallway	122
Oriental visitor	128
Five days from Selby	132
Standing under the lee of a single post	156
Come over here, Dick! It's deeper!	184
Always reminded me of the seven dwarfs	193

Photographs courtesy of W.H. Young and The Evening Gazette.

All drawings by the author.

Cover Illustration adapted by the author from an original drawing by Sam A. Roberts, 23.4.32.

INTRODUCTION

This is the story of a Boathand Apprenticeship served by young men in the Tees Pilotage Service just after the Second World War. All the incidents are true and any inaccuracies are only in minor detail. The characters and situations are as they existed then and only the names have been slightly altered to avoid any possible embarrassment. It is intended as a tribute to a way of life and a kind of seamanship no longer practised locally and so in a way is historical although very much within living memory. It is hoped that the reader may gain some insight into the way things were done and may enjoy the humour in the situations which arose at the time. It is written about a time before modern technology as it is taken for granted today, about a time before the Tees developed into the major port it is now, and when Pilotage was still very much a tightly knit, almost family, occupation, with all of its problems and implications.

CHAPTER 1

The carriageway of the Transporter Bridge grumbled and groaned under our feet as we stood on the very top span of the local landmark, a good spot from which to view the twisting river, coffee coloured from the muddy water brought down by recent rains. Five young lads, fresh from school, had been sent from the Pilot Office to the riverside to take a look aboard the steam Pilot cutter that was somewhere between the coal bunkering hoist and Corporation Quay near the Transporter.

The bunch of us had been told we were successful candidates for a job as Pilot Boathand Apprentices and the steam cutter was to be our place of employment for the next four years. We were all keen and eager for adventure. Little did we know!

Arriving, chattering, at the quayside, there was no sign of anything resembling a Pilot cutter so Geoff Mulligan suggested climbing the one hundred and eighty feet or so onto the walkway on the top span for a good look around. It seemed a good idea, so up we went!

Geoff, whose elder brother had already served his cutter time and had gone to sea, looked around below us and then suddenly pointed about a mile downstream.

'Here she comes,' he yelled, and we spun round to see a tiny nondescript craft rounding a bend from behind a high slag embankment. Its slightly battered, down-at-heel appearance was accentuated by the clouds of coal-dust that seemed to swirl about from it at each gust of wind. Black smoke poured from its dirty funnel, matching the heaps of coal lying scattered about its decks and superstructure. It had the appearance of a crawling beetle that some naughty child had poured muck on as a torment. Alan Hinton spoke first.

'Crikey, what a mess.' His voice conveyed our thoughts as the vessel approached us up river and passed beneath our feet. Our first glimpse of the *B.O. Davies*, as she was called, was a unique, bird's eye view.

The wake suddenly altered direction as the helm was applied and the little craft began to swing around to port.

An anchor chain rattled out of a worn hause pipe leaving a stream of creamy bubbles on the soupy water, and the rust-streaked hull swung gently round to stem the sluggish flood tide.

A wheelhouse window dropped with a bang.

'On deck!' We could clearly hear the bellow even at our altitude. A stubbly face appeared at the open window. Captain Jack had had a hard night steaming around the entrance and his temper was not sweetened by a full day bunkering and taking water and the prospects of pretty much the same tonight.

'On deck!' This time his voice had a tinge of aggravation in it. It did not go unheeded. A scamper of feet along the deck and two urchins appeared dressed in a motley assortment of old clothes. These two joined the third who had let go the port anchor.

'Stand by with a fender.' Jack's head popped back into the wheelhouse and he swung his wheel and rattled the telegraph simultaneously.

For a moment there was comparative peace as the little vessel nosed in towards the wooden jetty, a brief flurry of water under her stern as the engines were reversed. The bow barely brushed the quay as one lad leapt ashore to drop the eye of a mooring rope over the iron bollard. In a trice she was moored fore and aft and the boathands busied themselves arranging a long canvas hose into the main water tank on the boatdeck.

Jack Emmerson rang 'finished with engines' on the telegraph, slammed the wheelhouse door and clattered down the brass-bound bridge ladder to disappear below. The urchins on deck gave the moorings a final check, screwed up the brake on the anchor and scuttled along to the galley before the engineer or fireman grabbed the frying pan. While the water hissed into the big tank it was time to eat.

Seeing all this from our high perch we looked at each other. It was Doug this time who broke the silence.

'Well, come on, then, let's get down there and have a look.'

Down we clattered on the steep, steel ladders until we arrived at the bottom, our hands and clothes filthy from the grime of the bridge. Almost timidly we approached the cutter and it all seemed much larger and more chaotic on closer inspection. Coal dust smothered everything in sight. Great heaps of the stuff lay about the decks and abaft the salt-streaked funnel. The fireman leant against his shovel and wiped his grimy face with an even grimier sweat rag. He grinned at us and beckoned us aboard. Cautiously we stepped onto the boat deck straight from the level of the quay and, as we had been told to introduce ourselves to the Captain, we jostled for position to be last down that ladder.

Once aboard we were in a completely different world.

The general impression was one of staunch, well used, almost tatty purposefulness. No frills, bare painted steel, and everywhere that smell of steamers, hot oil and steam pipes, salt water-soaked gear drying on hot, dripping radiators that give a convulsive little thump from time to time. Coal dust everywhere. The only touch of elegance was faded teak doors into the saloon and gleaming brasswork all around. As it was almost low water, the cutter lay below the level of the towering quay with a kind of quiet serenity that only steam-powered craft have. No roaring diesel generators or auxiliaries, just the barely discernable pulse of the steam genny and the quiet but all pervading hiss of the dormant engine room. All the accommodation lay in the shadow of the quay which became almost a darkness as we went below to find the Captain.

Jack Emmerson sat alone in his tiny cabin, a pair of tired feet up on the desk top, his battered uniform cap pushed back off a weathered brow and an enormous pot of black, strong tea lying within reach. He was ruminating on the day's events.

The sudden arrival of a bunch of fresh-faced teenagers down the narrow stairs disturbed his thoughts and he looked up as the first of us arrived, blinking, at his open door.

'Now, my lads!' he boomed in a voice tinged with fatigue and lacking in enthusiasm.

'We've been sent to look around and report to you,' blurted Geoff, as we all jostled around, filling the tiny doorway.

Jack politely asked us each our names in turn and, after a long pause, said, 'Go on deck and find the Senior Boathand. He'll show you round when he's had his dinner. Stay out of the way and you can come for a trip down the river when we leave if you like. We're only takin' water – it won't take long.'

We were on our way back up into the daylight when the skipper called Geoff back to his room. We waited for him in the tiny cross alleyway at the top of the staircase as he questioned Geoff about his elder brother. This cross alleyway opened out on deck at either side and was the only access to the main saloon used by the Pilots. It was the grandest part of the little vessel, being panelled in polished mahogany and all the fittings being of cast or polished brass. Two large tables lay parallel fore and aft and a buttoned settee curved right round the for'ard bulkhead. Mahogany easy chairs, chained and screwed to the deck, were upholstered in blue moquette to match the settee. Curtains at the polished brass ports, brass lamps and a brass railed cupboard completed the furnishings. The only odd thing about it, and it gave away the fact that the vessel was so small, was the pronounced up-hill walk as you entered the doorway. We soon found out later, when we had to scrub the floor, that you had to start at the top of the slope, otherwise your dirty, soapy water ran back over your clean floor.

Almost directly opposite the saloon door on the other side of the cross alleyway was the serving hatch into the galley. This had a small window in it so as to see what chance you had of cooking a meal! A serving hatch has a slightly grandiose ring to it and conjures up a picture of white-coated stewards dancing attendance on a subtly lit dining saloon, silver tureens and cutlery glinting on white, starched linen in the soft gleam of well trimmed lamps. Not a bit of it. It had a purely practical purpose.

The story goes that a very senior Pilot and Cutter Company shareholder was cooking his breakfast one winter's morning. His wife had sent him out with a lovely gammon steak, mushrooms, tomatoes, two fresh eggs and even a ring of pineapple. He cleared the galley, scalded out the frying pan, put a knob of pure dripping in the bottom and began to cook with mounting anticipation. The cutter lay at anchor, head to wind, as the rain whistled past the galley door. Inside, the old veteran warmed a plate and carefully transferred everything from the black skillet onto it, picked up his knife and fork, swung open the galley door and, stepping outside, made for the cross alleyway door. A sudden gust of wind tilted his plate and his gammon steak with the fried eggs riding on top went sailing over the side and straight down the gullet of an amazed gull that happened to be paddling alongside in the hope of a windfall. I don't know who was the most surprised, the delighted gull or the irate Horace. Anyway, after that, Horace always used the serving hatch.

Meanwhile, Geoff had rejoined us and, passing the open hatch, we heard voices. Through in the galley the Chief and his fireman were busy preparing some culinary delight. The engineer was wrestling with something in the vast, pot sink, while behind him the fireman was throwing things into a large, iron pan, bubbling away on the stove. Both seemed unaware of our presence.

'Is the kettle boiling, Chief?' a polite enough question.

The Engineer, a little man, bald and with the hooked nose and prominent chin of Mr Punch, muttered some obscenity to the effect that it was not and what's more wouldn't be as it had just been filled with cold water.

Unabashed, the fireman, who had two pint pots in his hand with tea leaves, condensed milk and a mountain of sugar in each one, grabbed the pan boiling the potatoes, and expertly slopped the required amount of water into each pot, grinning from ear to ear as he did so. Mr Punch never said a word but took his proffered share, stirred it vigorously, and leant back against the warm, steel bulkhead that formed the galley wall. I gathered that their lunch would be some time.

We innocents picked our way along a deck strewn with coal, buckets, trimming shovels, deck brushes and an assortment of other items which had been abandoned for the moment. Down below a feed pump banged monotonously away, unheeded in the depths of the engine room.

Peter almost fell down a manhole left open in the deck and effectively camouflaged by a cunningly placed heap of coal. It was the entrance to a side bunker in which someone had been working when they came alongside. We all gawped in at the door to the fiddly and down through the gratings to the stokehold below. The top of the boiler curved away into the darkness, covered in very fine, grey ash. The rails were draped with boiler suits and other items drying in the dusty atmosphere and everywhere the soft, pervading hiss of captive steam. Many times in later years, I came to appreciate the dry, dusty warmth of the fiddly top as we crashed and banged about the North Sea in winter time. It was somewhere to keep dry and warm, out of the way without being accused of hiding.

Passing the companionway up to the wheelhouse and boatdeck we eventually arrived on the fore deck. Here was the fo'castle companionway leading down below decks, the entrance protected by a curved steel scuttle and a pair of narrow teak doors, usually permanently hooked back.

Our leader peeped down into the gloom.

'Anyone there?'

Something stirred below.

'Wadya want?'

'We are the new boathands,' said Alan politely. There was a pause, then, 'Thank God for that. Come on down if you can get down.'

We all half climbed and half fell down the steepest set of iron-bound stairs you've ever seen into a strangely shaped compartment with painted steel bulkheads.

Arranged against the aft bulkhead, the only wooden one, were four iron cots or bunks in two tiers. Only three had mattresses; the fourth carried a pile of ripped and tatty oilskins, an old submarine sweater and a couple of ex-army haversacks. Thick brown linoleum covered the floor and the table top and arranged along the port side was a plastic-covered settee. A set of battered wooden lockers and a wooden form completed the furnishings.

This was where the boathands lived. The riveted steel plates ran with condensation as most of the fo'castle was at or below the waterline. The beam knees and angle iron stiffeners were left bare and unadorned as indeed were the rows and rows of rivet heads that marched up the bulkheads in orderly ranks. No mahogany or moquette here. The galvanised portholes each with their deadlights latched horizontal let in a reluctant watery daylight that danced reflections on the painted deckhead.

The place was in utter chaos. Sea boots scattered and flung in various odd corners. Old army blankets in disarray on bunks and settee, a couple of grubby towels drying on a tiny iron radiator which dripped into an old pudding basin. A mirror strategically placed, presumably so that you could make sure you looked your best before appearing back on deck. Stan, the senior hand, gave us a cool, slightly disapproving look.

'Thank Christ they are going to start someone at last,' he repeated. 'Maybe now we'll not have to work twenty-fours.'

He explained patiently that because several of the senior apprentices had recently left for sea service, those remaining were having to work twenty-four hours on and twenty-four off 'and on this thing,' he added, 'it isn't funny after a couple of weeks!' We didn't have time to press the point as a door up top slammed open and a familiar voice announced, 'Tanks full, lads. Stand by!'

The smallest of the three boathands who had been on the top bunk 'doing a deckhead survey' suddenly leapt off the bunk, dodged round the pillar stanchion and pounded up the companion ladder in a series of well practised leaps, disappearing into the sunlight.

Stan said we could keep out of the way best on the after deck and to watch what we touched as everything was filthy until they managed to get hosed down.

Once again the telegraph rattled, lines were cast off and a prodigious blast blown on the yard-high, brass steam whistle, to warn the Transporter Bridge of our departure. The *'Old B.O.'* swung her bluff bows into the flood tide and with the uncanny, almost silence, of a steamer, glided off down the river. We newcomers arranged ourselves right aft on the towing rail to survey a scene which, although completely new and full of interest now, was to become very familiar in the months and years ahead. The grizzled skipper stood alone in his wheelhouse whilst his crew toiled to get the craft tidy and ship-shape. The last remaining coal was shoved down the holes. Cross bunker hatches were battened down, decks cleared of clutter as the salt water hose was rigged. At this point the fireman went below to start the pump at the request of 'water on deck' and the boathands 'turned to' to slosh and splash and scrape away the last remnants of coal dust off bulkhead and bulwark into the scuppers and eventually over the side.

As this process was going on, the true shape and form of the old steam cutter began to emerge. Nobody would have called her a handsome vessel, her proportions left something to be desired, but then she had not been designed as a rich man's toy or a prestige symbol. She had been conceived and constructed, towards the end of the depression days of the thirties, as a cruising Pilot cutter. It was said she had been built mainly with apprentice labour out of what they could find lying about in the shipyard. Certainly she had been very heavily built to withstand hard and constant use and some of her fittings, such as bollards and fairleads, had clearly been destined for a far larger ship. I think her main drawback was that she was under-powered and I suspect her modest triple expansion steam engine was installed with an eye to economy rather than performance. She may have functioned adequately when new, but when we joined her she had just returned from war service as an examination vessel. She was a little past her best and it showed, when boiler cleaning time came round and the grass and barnacles grew long and profuse below the waterline. Succeeding generations of boathands and others worked hard to keep her youth as long as possible, but the price of coal and wages, together with ever higher maintenance charges, eventually sealed her fate. She became too old and uneconomical.

However, painted as she was in sober colours, the deck hose revealed her secrets one by one. Her all-riveted hull was black with TEES PILOTS in large white letters on each side. Her superstructure and funnel were buff, the latter having a black top and bonnet. The winch, davits and windlass were all bright green with the ends of the windlass barrels painted in roundels of white and red. This latter touch of flamboyance was directly attributed to Jack Emmerson's days with a local tramp company (true or false it was certainly in character and also

Ropner's colours).

Handsome she might not have been but she was certainly full of character in a way that steam-powered craft often are. The steamer's engine room seems so much more alive than the roaring, smelly, teutonically efficient diesel. Dammit, you can see what's going on; you can see the piston rods pumping away; you can see the connecting rods transmitting the power and the crankshaft flailing round. It's the comparison of steam locomotive and Deltic diesel. The diesel is just a lump on wheels but a steam locomotive like the 'Cock-o-the-North' was a living, vibrant bundle of energy. Small wonder people become attached to them. Seafarers, in particular, often have strong views on ships they sail in. A ship is much more of a living, moving thing, not confined to straight and level rails. It moves in all directions. It reacts to its surroundings all the time. It is in conflict with the elements. It responds to human will and is at one time a home and a haven from the storm. Not surprisingly it generates a certain amount of affection. Its sounds become familiar as they are lived with, sometimes for years – the creaks and groans as they roll to swell, the suck and gurgle as the water in the bilges moves about. The clack and shiver of the anchor cable. The slap of the halyards against the mast as the wind moans through the canvas dodgers. Last but not least the attention they demand finally puts them firmly in the female gender. There can be no other explanation for it.

Coal-fired steamers are the workhouses of the sea. Ships never needed so much constant attention and dedication since the days of sail. Prolonged effort is needed to keep them vital and alive. Coal has to be loaded and trimmed so as to pack the maximum tonnage into the minimum space. Once bunkered the fuel has to be shovelled into the furnaces in the stokehold where a constant watch has to be kept on water level and steam pressure. Every day the fires eventually become choked with clinker and ash and have to be cleaned out. The enormous, twelve foot long poker or slice breaks up the lumps of slag or clinker and the equally heavy rake pulls all the unwanted clinker and cinder out onto the stokehold deck plates where it is next loaded into great iron buckets, three feet tall, and hauled up the lift by hand to be finally dumped over the side. A back-breaking job for the fireman. Of course, if the fireman had failed to turn out for one reason or another, it was left to the boathands to make themselves useful in the stokehold. We were nothing if not versatile; in fact, our greatest value lay in this versatility. Pilot Boathand Apprentices might be our title but we acquired the skills of a variety of trades including painter, signaller, steward, rigger, able seaman, fisherman, fireman, mechanic, char and maybe others I've forgotten.

This was what we were letting ourselves in for. This was the life we had chosen and we were happy to be accepted into the Service, and this was the '*Old B.O.*' as she was affectionately called by everyone. As she steamed down the Tees on that fine May afternoon in 1947 to resume her sea station, it was as if to meet a brave, new, post-war world. Food was rationed as rigorously as ever; there was only one grade of petrol in the pumps; Sid Field's new film *London Town* was playing to packed houses at 1/3 and 1/9 a time and Teresa Brewer's song: 'Put another nickle in the nickleodeon' was in the top twenty.

If someone had come along and told us that in less than thirty years the Tees would be one of the major oil exporting ports in the world we'd have said 'Don't be daft.' If he had gone on to say that the crude oil being shipped out was coming through a giant pipe line laid on the sea bed from enormous oil drilling platforms sitting two hundred miles out in the North Sea and that this was only one of the many North Sea oilfields that were producing enough oil for the UK to be self-sufficient, we'd have said, 'You're barmy, mate.' Science fiction indeed!

CHAPTER 2

The waiting room of Bodger's Taxis at Redcar Station is a very modest affair. Old 'bus seats are arranged to give some degree of comfort while the customer awaits his turn.

As the train from Marske trundled in and deposited its passengers, several elderly gentlemen appeared almost simultaneously in the vicinity of Station Road and converged on the tiny taxi firm. Most of these characters were dressed more or less in dark blue uniforms and wore the white topped uniform cap which, in those days, proclaimed the summer solstice. The night watch were preparing to relieve those that had been all day at the breakwater.

In the waiting room sat Peter Jackson, a slightly nervous, chubby cheeked, tousled haired lad of not yet sixteen. He had arrived on the train from Marske and had a brand new haversack slung over the shoulder of his brand new oilskin. His white socks and wellies were also new and on his head was a knitted woolly hat. This was his very first watch and eight o'clock in the morning seemed a long way off! He sat alone and fidgeted on the bouncy upholstery of the seats. Suddenly the door burst open and Jack Nixon appeared. He had the appearance of one who had dressed in a great hurry. The white top on his battered cap gaped at the seam, his faded blue uniform mac had two buttons missing and the belt was fastened out of the way at the back; his crumpled pants were at half mast revealing heavy knitted grey socks and brown boots with broken laces. His morning shave, apparently cursory and abandoned, had not been a success and little bits of tissue were still stuck on his face to staunch the bleeding. His whole appearance was wild and unkempt and it was a startled Peter he addressed.

'Now, sonny, are you one of these 'ere new lads?'

'Yes, sir,' Peter replied. 'This is my very first watch, sir.'

The elderly Pilot regarded him with slightly bloodshot eyes and blew his nose noisily on a brilliant red and white hanky.

'And 'ave yer just left school, then, sonny?' he enquired again.

'Yes, sir, last week, sir.'

By now two others of the watch had arrived to join the little group: one, a big man with florid complexion and slightly protruding eyes and another much smaller, swarthy man of about thirty. They all stood listening with some amusement while Jack Nixon quizzed the new lad.

'And what school might that have been, then, sonny?' continued the unkempt old veteran, by now warming to the interrogation.

'Coatham Grammar School, sir, just here round the corner.'

'Ah, yes, I know it well.' Jack Nixon thrust his hands in his trouser pocket and, leaning well forward as if to emphasise a point, he continued, 'and d'you know what school my lads went to, then?'

Peter began to sense an air of mild hostility and squirmed on his seat as he smiled back at the old man.

'No, sir.'

'Well, my lads went to Zetland School, sonny, and what's more, I'm the Mayor o' Redcar – what d'ya think of that!'

Peter, being a polite lad and anxious to please, thought this was his cue to respond to the old man's sense of humour, and laughed at the joke. The old man's face blackened and Peter realized too late his colossal *faux pas.*

'Oh – so you find it funny, do you?' thundered the irate Alderman.

'Oh no, sir,' flustered the luckless lad. 'I just didn't think ordinary working men ever got to be Mayor.'

'Working men! Ordinary working men!' Jack's jowls flapped puce and purple as he coughed and spluttered, whisking out his enormous handkerchief again.

The situation was saved by the appearance of three more of the watch off the Middlesborough train and they all tried placating the injured pride of the local dignitary, who seemed in imminent danger of throwing a wobbler.

Peter, by now realizing that maybe he'd said too much too hastily, was bundled into the very back seat of a large, ex-army Humber shooting brake, while his elderly adversary was given the place of honour next to the driver. Peter Lipton, the senior lad, came running up the road at the last minute, clutching a *Sport's Gazette* and, with a great revving of engine and a cloud of blue smoke, the whole show took off with a great screeching leap as the balloon tyres clawed for traction on the oily, smooth tarmac outside the taxi stand. The relief was on its way.

A mile down the road the adventure was about to start for me. I sat on a seat by the roadside at the Coatham roundabout waiting the arrival of the night watch's taxi, oblivious of all that had transpired. I had just walked the mile and a half from home in heavy boots and oilskin and this, together with a large haversack packed with everything my parents insisted I needed, had caused me considerable discomfort. Coatham church clock reached and passed the appointed time and I began to get uneasy. Had they forgotten me? Even worse, had I missed them? Thirteen minutes to five – twelve minutes to five, I began to feel panic rising – then suddenly a squat, box shaped vehicle, swaying a little unsteadily, rounded the corner. It scraped to a halt and a door was flung open.

'Gerrin, son,' a rough voice emanated from the pile of humanity within. The only space that I could see was next to an old chap sitting with the driver! I squashed in, hauling my heavy bag after me. I thought I heard a muttered remark about 'Bloody kids' – the door slammed shut, trapping the skirt of my new oilskin, and we were off.

The journey to the breakwater was uneventful and conducted in comparative silence except for one of the Pilots in the back winding a window down and shouting some cheerful obscenity to the gatekeeper at the railway crossing as we passed over the lines. He responded with a lump of coal hurled in our general direction, as we lurched and rolled on our way.

We eventually arrived at the top of the jetty and virtually fell out in a crumpled heap. The steam cutter lay alongside the end of the jetty in the evening sunshine and we picked our way past the lines of hopeful fishermen and small boys to be met by a slightly pompous, middle-aged man dressed in a grubby white duffle coat and a black beret. He was smoking a cigarette in a long black holder and drinking a pint of tea from a clean, white pot. His shoulders had a pronounced stoop and he had that strange habit of closing his eyes whenever he addressed you face to face. Captain Hainsworth, to give him his full title, was for some reason always referred to as Blood. It was our very first meeting with him. We had introduced ourselves to Captain Emmerson weeks before, but had never encountered Captain Blood.

The day watch walked ashore and Alan Hinton and Doug, the other new starters, raced up the jetty, pulling on coats and trailing haversacks as they ran. For a brief moment there was confusion as the watches handed over in haste and then, with the customary cloud of blue smoke, they were gone. Peter Lipton, our Senior Boathand, had six months experience and so was an old hand. It was his job to lick us into shape as quickly as possible and he began by making it quite clear to us that he had scrubbed his last floor and possibly cleaned his last brasswork. Ours was a very subordinate role, at least for the foreseeable future. Captain Blood briefed us on what he expected from us and told us to be at all times respectful to the Pilots we encountered and that he would not on any account tolerate the use of bad language on deck or anywhere else for that matter. I remember thinking this rather amusing as the only bad language I had heard so far had been from the Pilots themselves. Some seemed to use it instead of punctuation, even managing to insert a profanity between syllables of a word. But ours was not to reason why.

We stowed our gear down for'ard and had our tea, discussing all the time the whys and wherefores of the night ahead.

Peter, the senior, said he would have to go in the boat every time we had a ship to attend to at least until we were familiar with the geography of the place and could be trusted to find our way about without hitting the bottom. It all sounded very strange and very exciting.

We were told, in order to avoid confusion and sarcasm, always to use nautical terms. For instance, you never go downstairs, you go 'below'. Walls are 'bulkheads' and ceilings 'deckheads'. The floor is 'the deck', and a window is a 'port'. You don't tie up, you 'make fast' and if you're clumsy the line doesn't snap, it 'parts'. You come ahead and you go astern and, contrary to whatever the Americans say, port is left and starboard is right. We tried very hard not to drop any clangers and eventually we began to sound like sailors. Of course, when Captain Blood bawled at us to 'vast heaving and belay' we knew we had really arrived!

After tea, it was our first job to collect all the dirty crockery and wash and dry them in the galley sink. We managed this with a good deal of leg pulling, Peter and I doing all the work while the senior Peter went below to the summons of Captain Blood in his cabin. Pilots came and went in and out of the saloon and, from time to time, small boys standing on the jetty (as we were still alongside) asked us when we were leaving. Apparently the presence of the cutter was upsetting their fishing.

At that moment a very distinctive sound reverberated in the air of the summer's evening – MMMBBAAAWWW – It was like the amplified croak of a giant bullfrog – MMBAAW – MMBAAW – MMBAAW. The sound tailed away in the still air, for a brief instant echoed back from some distant point and then silence.

Peter, the senior lad, came pounding up the stairs from below, shouted for me to follow him, and, leaping on to the jetty, made towards the boat that was moored to an iron ladder round the other side at the back of the piles. As he swung the starting handle, the engine spluttered into life. He signalled me to let go the painter, the painter being the short length of rope spliced to an iron ring in the fore part of the boat and used in mooring when no boat rope is available.

He swung the boat round the corner of the jetty and roared out from behind the cutter's rusty bow straight into the burning glow of the late evening sun.

The water was a sheet of burnished brass. It was almost a shame to disturb it. Our bow wave cut an angular pattern across the surface which reluctantly faded away several hundred yards astern of us.

Peter, the cock hand, explained that we had been waiting alongside for this ship to come out of the river and the signal blown on the whistle, one long and three short, meant the Pilot on board was ready to be taken off. It was a collier or 'flat iron' as we called this particular type, because it was purpose built to travel up the Thames. They had to negotiate quite a lot of extremely low bridges and so were built to be as close to the water as possible. Masts were taken down and, once rigged for the passage through the bridges, they had all the style and charm of a tarry railway sleeper.

We turned and ran alongside, almost level with the for'ard well deck. Dirty water ran over the side from every scupper hole as the crew washed down the hatches and decks. Peter picked a spot where a pair of iron stanchion posts and a short ladder had been rigged clear of the cascading hoses. A leg appeared over the gunwale as the Pilot sat astride for an instant, checking that the boat was alongside. Then, with a wave to someone unseen, he climbed over and landed in the boat. A terse nod to us as Peter sheered off and we were heading back to the cutter alongside the jetty. Not a word was spoken as we ran back, the Pilot lost in his own thoughts, Peter looking where he was going and me trying to relate where we'd been to, where we were going, where the ship was going – seemingly a completely different direction – and what was going to happen next! It all looked so simple, easy and matter-of-fact.

Later on I learnt that we had taken a short cut in the boat, which entailed getting certain markers lined up in transit and sailing over a rocky outcrop with inches to spare under our keel! We tried the same thing one moon-lit night years later and hit the bottom with such an almighty crash it jarred our teeth. Someone had moved one of the markers!

Having landed the outward Pilot and seen him set off to walk the four miles or so to the railway station, it was announced that we would get underway from the jetty and proceed to our anchorage outside at the Fairway Buoy. To get 'underway' means the opposite to being moored or at anchor. It means you are free to move in the water. A ship can be 'underway but stopped' meaning it is lying motionless in the water not

Cutter Confusion

Stand by the anchor

attached to any structure, or it can be 'underway and making way' which means it is actually in motion.

Captain Blood appeared from below and climbed to his wheelhouse. A couple of the younger Pilots appeared on deck close by as we two new starters waited nervously to let go the mooring ropes. Bells jangled, water stirred under the little ship's stern, Peter, the senior lad, appeared on the jetty and threw off the moorings as the vessel swung to the ebb tide and cleared the end of the weathered piles. We steamed serenely off into the calm, deep water, pausing briefly to pick up the boat that the senior lad had returned alongside in, then we were off to sea in the quickening dusk of a perfect summer's evening.

We came to anchor about twenty-five minutes later. We were shown how to detect an anchor dragging on the bottom by feeling the vibrations through the links of the chain or cable as it is always referred to. Then, by looking over the bow and with one foot on the cable passing through the hause pipe, the cable is seen to come tight, stop vibrating, and then come a little slacker. At this point the vessel is said to have 'brought up' to her anchor, meaning that the process of anchoring has been executed successfully and the anchor is holding.

The anchor light, or riding light, was hoisted, plugged in and switched on together with the red and white all round lights on the mast high above the wheelhouse, indicating we were a pilot vessel on station at anchor.

The green and red side lights were switched off by the skipper as he rang 'Finished with Engines' on his telegraph. The cutter swung on her anchor, head to tide, with a final clack on her cable and we settled down for the night. A light breeze off the land barely rippled the surface of the water as the shore lights winked and blinked at us across the bay. The senior lad announced he would keep the first watch and call one of us about midnight. This was another of the perks of being top hand and one we later questioned, but to no avail.

The strangeness of it all, the unaccustomed noises, the odd cackle of laughter, the opening and slamming of doors, the hiss and thump of the radiator: I lay awake seemingly for hours. Eventually soft snores came from Pete in the top bunk, noises started to merge and drift into the pattern of a fabric of darkness. I was shaken firmly by the shoulder.

'Come on, your turn on watch.' Peter Lipton was sitting on the bench, his sea boots already off and obviously ready for the horizontal.

'What's the time?' I asked, fuddled with the unaccustomed awakening from a deep sleep.

'It's nearly one, and there's an East Asiatic Dane sailing at two, so keep your eye open for it about half past three. Give us a shout as soon as it blows for us. Keep your eyes open and don't let the galley fire go out.'

I climbed the steep ladder clutching my sandwich tin and other bits and pieces. The night was moonless and the land breeze had a softness to it that is quite rare in northern latitudes. The cutter hardly stirred in the calm and I walked the decks trying to orientate myself with the blackness of the northern horizon and the necklace of shore lights encircling the other half to the south.

Time drifted by, sometimes quickly in conversation with the fireman, sometimes slowly when my various chores were done and constant glances at the clock in the cross alleyway seemed to indicate little progress. Three o'clock came and went. Three thirty, and a faint pale streak appeared in the eastern sky.

A deep, rich, fruity horn sounded loud and long as the Danish passenger-cargo ship rounded the leading lights four miles away. The *Lalandia* was right on time. I went below to wake the other two and get into my oilskin as a pre-dawn chill was apparent in the air. The other two lads shivered as they came on deck, fresh from a warm bunk, but we soon had the boat in the water and were running into the breeze with the canvas hood standing against the cold while we warmed to our stiff oilskins.

The white painted mail boat was like a ghostly sailing ship as she glided towards us. Four steeply raked masts could easily have had yards on them and the illusion was made complete as she had no conventional funnel but merely a pair of thin exhausts running up alongside her mainmast. Rows of large windows on her promenade deck, with the occasional glimpse of a glittering interior, gave her a floating, fairy palace appearance. She slid towards us with the faintest of murmured burbles from her exhausts. The sight of one or two hardy passengers lining the rail, muffled against the morning air, waiting to see the Pilot leave, suddenly made us feel that our job was of interest to others; we tried to look efficient and nonchalant at one and the same time.

Peter Lipton laid the tiny cockleshell alongside the towering white hull; the steel plate edges with their rows of rivets cut little eddies in the smooth water alongside. An immaculately dressed officer inspected the ladder. I remember thinking how alert and efficient he looked, clean shaven, starched white shirt, freshly laundered cap cover, gold braid gleaming in the light of the cluster of bulbs illuminating the ship's side. He must have been up most of the night also – maybe he was used to it! I suddenly started to feel very weary. The night had been a quiet one, hardly any shipping at all. I reflected, did a busy time make you more tired or less, or was it simply the lack of undisturbed sleep that I wasn't used to? Although daylight was by now firmly established in the eastern sky, the sun was not yet awake and eight o'clock relief time seemed a long way off. My thoughts were disturbed by the Pilot's arrival in the boat. A slim, almost skinny, man with a beak of a nose and a peppery look about him, he turned and waved up to a duffle-coated figure on the bridgewing and in a thin, quavery voice shouted, 'All clear', though I doubt anyone heard him. The figure waved and disappeared into the wheelhouse. Almost immediately we heard the puff of compressed air as the engines were re-started to a subdued chunter. The passengers drifted away from the rail to the warmth of their cabins, the cluster of lights were extinguished and the great, pale ship swung onto her new course and gathered speed.

Ginger Billy sat on the engine box and regarded me from beneath great, bushy eyebrows. Suddenly he spoke in a high pitched falsetto which seemed to cause a vibration period in his front dentures.

'Are you the new lad then – EH!'

I confirmed the fairly obvious and he went on to ask my name and how I was liking the job so far. All the time his front teeth fluttered away furiously, though he seemed oblivious to them. Remembering Peter's experience the night before, I was careful to say nothing controversial as I was too tired to be drawn into a confrontation and, after all, for all I knew, he might have been the Mayor of Middlesborough!

CHAPTER 3

The Cutter Company was run on a shoe string. Never was there a penny spent if a ha'penny would do. One or two of the oldest Pilots still had shares in the company which they had bought when they were young men and the way they pinched and scraped gave the impression that they hoped some day to be paid a dividend.

The skippers were obviously instructed to make the stores stretch. They issued paint or rope or even soap for scrubbing out like giving blood. We of course, at the time, didn't realise the reason for all this, we just got the thick end of Jack Emmerson's temper if we weren't careful. To leave soap in water to go soft and waste was a heinous crime. As long as a rope hadn't to run through a block it had to be spliced one more time. Brasso was liquid gold and Vim scouring powder a pure luxury.

For whatever reason we lads had for wanting to serve an apprenticeship, it certainly wasn't for the money. We started off at a fraction over a pound a week for an average fifty-two hours! Our apprenticeship was a strange one-sided affair in so far as we were guaranteed nothing. If we completed our four years service we were sent deep sea to obtain enough time to qualify us to sit for our tickets, a common term used for Certificates of Competency for Master and Mates.

You could then apply to be considered for a Pilot's Licence. If there was no vacancy you had to wait. If you failed your tickets you could not apply. If your health or, even more important, your eyesight became defective, you were not considered. If you became too old, your chance had gone. In retrospect, I shudder to think of all those dice that were loaded against us. Notwithstanding, we accepted it all with a light heart and the optimism of youth. Of course, because of the abysmal pay rates none of it would have been possible without parental help. So, I suppose, it could be argued that the lads' parents subsidised the Cutter Company.

The five of us who started together were split into three watches. We worked continuously, having no weekends or Bank holidays. Sometimes, if someone failed to turn up, you had to stay and work a double shift. Very often, especially in the wintertime, we had late reliefs and had to work an extra hour or two. No overtime was paid for this; it was expected as part of the job. Sometimes, when dense fog descended on us, we were unable to move and the night watch didn't get relieved until mid-morning. As one engineer put it, 'On this job you're expected to work Chinese watches – come on and stay on!'

Our duties were many and varied but our basic function was as crew for the cutter, keeping it clean, working as stewards to the watch of Pilots currently on board. When weather and shipping permitted we chipped and scraped her rusty sides, then red leaded and painted, or else we washed down paintwork or 'soogied' as it was called. We picked up the anchor and dropped the anchor sometimes a dozen times a day. We manned the boarding boats every time a ship required a Pilot or a Pilot wanted landing. This was really our main function and in good weather we revelled in it. We soon became very expert in the dropping and picking up of the little motor boat. When the weather was fine and the steam cutter remained at anchor on station, we were left very much to our own devices, often spending many hours in daylight or darkness just motoring around the North Sea attending shipping as it arrived or departed from the estuary and occasionally putting outward Pilots ashore for them to walk the few miles to the nearest public transport. Very few, I think six at most, owned a car.

Of course the boatwork was our first love because it gave us a great opportunity to exercise our initiative. In fine weather it was to be enjoyed; in wind, rain or just plain bad weather, it was to be endured and we cursed it. We lay about on the side benches in shirt sleeves, heavy trousers and sea boots, enjoying the occasional fine day or we huddled behind our rough canvas hood as sea after sea broke over us, one lad at the tiller wearing two oilskins, one on top of the other while his pal sat on the engine box pumping furiously at the bilge pump as the boat slowly filled with water. We had to keep the water level below the propeller shaft coupling or else the water sloshing about would be picked up by the spinning flanges and we got it up our knicker leg. Nothing dampens the spirit more than a wet crutch.

It was, however, this recurring obsession for economy, coupled with a keen desire to avoid arousing the skipper's wrath, that perpetuated one of the funniest incidents that happened early in our boathand days. It all began one morning in the early summer. It was a 'scrubbing out' day, as Mondays, Wednesdays and Saturdays were known, and Geoff Mulligan was preparing hot water on the galley stove while looking around for his favourite bucket. Geoff had an uncanny knack of rubbing the skipper the wrong way and, try as he might, he always seemed to put his foot in it. Jack Emmerson's temper, never one of his more endearing traits, seemed to be on an ultra-short fuse where Geoff was concerned. He didn't suffer fools lightly and, let's face it, at sixteen, Geoff could be a bit of a fool. The cutter was due to go for bunkers and at the same time collect stores from the chandler's wagon at the coal hoist. Captain Jack called Geoff down below to his tiny cabin and gave him the last chunk of soap in his stores' locker with the clear message that he expected a thorough job doing, make the soap last out and no skimping underneath the bunks.

In solemn silence Geoff took the last bit with almost an air of reverence and carried it up on deck as though he'd been entrusted with the Olympic Flame, only knowing Geoff's luck the thing would have blown out going round the first corner and he wouldn't have had two pence for a box of matches! He filled his bucket, selected the scrubber with the longest bristles – there were only two – and began the tedious chore that sometimes took a couple of hours.

First of all the ancient coconut matting runners had to be rolled up and carried out on deck, then the whole of the linoleumed floor had to be swept clean of dirt and debris. Then off with your sweater, up with your shirt sleeves and down on your knees. If you could scrounge a bit of old towelling to kneel on it made things more pleasant, but the crawling and reaching under bunks in far corners soon had you sweating and sticky. The hot, soapy water soon turned hard calloused hands soft and white and crinkled your finger ends. Perspiration ran down your brow and dripped off the end of your nose with the exertions of scrubbing. Geoff slaved away in the gloom down below while the sun sparkled on the water outside. As his bucket of water turned progressively milky white then darker and murkier, the scum floated to the surface as he wrung out his floor cloth for the umpteenth time.

Jack Emmerson sat in his cabin, his feet up on his desk, reading the previous Sunday's *News of the World* and feigning lack of interest in Geoff's labours. As the luckless lad scrubbed round the legs of his chair, giving the soapy lather his best efforts, Jack spoke suddenly, causing Geoff to bang his head underneath the seat of the chair.

'I hope you're not planning to use that mucky water in my place, Mulligan?'

'Oh no, skip,' lied Geoff. 'Just going to change it now. I've got a panful of fresh on the stove.'

'I should hope so, but sometimes you seem that gormless I'm beginning to have me doubts about you. Go and change it now and hurry up about it. You've got all the saloon and stairs to do yet.'

'Aye, aye, skip,' Geoff flustered, rubbing the rising lump on his head. He grabbed the heavy galvanised bucket and staggered up the narrow stairs, slopping the filthy water here and there on the way.

Arriving on deck out of the cross alleyway, with a great effort, he heaved the contents through the bulwark door straight over the side into the sea, narrowly missing the boat that was moored alongside!

In the instant the water had left the bucket he realised his blunder! There in the middle of an ever-spreading patch of filth in the surrounding clear, blue water was his floor cloth – and the only piece of soap for miles – slowly sinking!

Geoff hit the water with a classic racing dive that Weissmuller would have been proud of. Someone shouted, 'Man Overboard!' but by the time the watch arrived on deck, the dripping Geoff was back sitting in the boat, grinning from ear to ear, clutching his slippery prize!

Down below, Captain Jack, seeing everything through his porthole just above the waterline, had a quiet chuckle to himself and resumed reading the *Barmaids' Weekly*.

'Well,' he thought to himself, 'If I go up on deck, I'll only have to bawl him out.'

This he often did for at least the first year and we all lived in mortal terror of Jack's temper. It was a deliberate policy practised by him to lick us into shape quickly and it seemed to pay off. We didn't make nearly so many mistakes in the second year. He even mellowed towards Geoff eventually.

Another of our many duties was keeping the Pilots' saloon clean and tidy, washing all their dirty crockery, making the bunks in their sleeping berth and generally dancing attendance on them.

With six or seven grown men gathered together, often for hours on end, in a confined saloon, free to do what they liked, to be as untidy and vociferous as they wished, this was no mean task. Remnants of meals were left abandoned everywhere. Ashtrays were provided but largely ignored. Items of clothing were scattered and intermingled with old newspapers. One gentleman spent his time making salmon nets, fathoms and fathoms of it draped like curtains everywhere, over lamp brackets, chairs, coathooks and often spreading over half the floor. A card school was often the most popular form of entertainment and this could become rather lively, especially if a bottle appeared on the table and the kitty was mounting up. Solo, the most popular game, was played with intense concentration and with three or four pounds in the pot, tension grew with the successive hands.

One summer's night, the weather was clock calm and when the daylight faded and the lights ashore shone bright, their reflection in the water had that long, still look about them that was a sure indication of fog to come. The night was going to be long and lonely. The watch of Pilots, with nothing much to do, began playing cards about nine o'clock.

Each watch of Pilots, and there were six in all, had distinctive and separate patterns of behaviour. Some would study the form book and spend hours pouring over the racing papers. Others would fish avidly and spend a whole night tending lines or gutting and filleting. Yet another watch would discuss politics, arguing the whys and wherefores until the small hours.

Number three watch's forte was cards and Solo in particular. They played for hour after hour and they played for money.

One or two small ships came and went during the course of the night but slowly, almost stealthily, the feathery fingers of wispy vapour silently drew a veil across the horizon. Shore lights fuzzed, dimmed, and were gone. By midnight it was solid thick. It was just possible to see the water from the boat deck.

The necessity to keep a visible lookout was gone. All that was needed was an ear cocked to hear the blast of a ship's whistle, moving in the gloom, or the clattering rattle of the bell of a ship at anchor. The all-enveloping murk cocooning the little vessel accentuated the shouts and excited voices of the card players as hands were lost and mistakes discussed at great length. The players sat round one of the tables, shirt sleeves rolled up, neck ties slackened or discarded, tobacco smoke lazily curling around the soft glow of low powered bulbs.

Ephy Dee constantly licked the tips of his fingers as he thumbed through his cards, swinging his short legs back and forth under his chair. Joe Braithwaite, balding and bespectacled, viewed his cards through a blue haze as he dragged away to keep his ancient briar going. Charlie Rawlinson, tall and slim, sat round-shouldered against the hard, straight back of the moquette settee. On his left Johnny Franks, youngest of the group, smoked his umpteenth cigarette, his strong, nicotine-stained teeth parted in a constant grimace. He badly wanted that pot that was growing with every hand. Brushing his lank, black hair from his brow he broke the back of the 'Camel' in the ash tray in a nervous gesture of anguish and then cursed; it was only half smoked! As John reached for the final pack of his favourite American cigarettes, Joe Braithwaite paused as he lit his pipe and started a tale about boathands frequenting the High Street pubs of Redcar on a Saturday night. Bill Bodger, the taxi proprietor, had met Joe in the Crown and Anchor and told a tale of teenage lads with fists full of notes drinking double rums as last orders were being called, prior to staggering across the Promenade and into the Pier Ballroom. This story had clearly made an impression on Joe as not only was he an aficionado of the High Street pubs but he had an only daughter who enjoyed her Saturday night dance. The idea of his only child being pawed over by a bunch of drunken youths made his hackles rise.

Now it just so happened that Douglas Buck, one of the names bandied about rightly or wrongly was, at that very instant, coming on watch at about four in the morning. He was unaware and completely innocent of the fact that the senior Pilot, and indeed a member of the examination board, was quietly nursing his grievance and waiting for a chance to confront the lad.

Every few minutes throughout the night as the card players shuffled and dealt, paused, considered, played, shuffled and dealt, footsteps could be heard patrolling the eerie deck outside, pausing to give the toneless bell a rattle and then fade away along the deck aft.

About six thirty there was a tap at the door, momentarily breaking the tension as the players turned to the intruder. The skipper, fresh from his bed, reported a slight improvement in the visibility and proposed taking advantage, picking up the anchor and sneaking slowly back inside before it clamped again. All the Pilots muttered approving noises and returned to the table littered with the debris of the all-night session.

The telegraph rattled overhead, its wires slapping and twanging their message to the engine room. The anchor slammed home in the hause pipe. A prodigious blast on the steam whistle and the cutter was underway, steering a course for the invisible lighthouse.

Once more a knock at the saloon door. This time the lad, Douglas, dust pan and brush in hand, peered around at the scene with obvious distaste. The saloon, looking like a rail disaster, had the atmosphere of a kipper shed. Four bleary-eyed men surveyed the table top covered by an old army blanket on which lay the best part of five quid in large and small change. Each searched his cards in desperation for the winning hand that would take the kitty before the eight o'clock relief.

Doug, cut short by another blast on the whistle, asked politely if he could start cleaning up. A grunted acknowledgement, and Doug began his chores. Suddenly Joe, thinking to give the proceedings a little light relief, turned on the unfortunate lad.

'Now, my young "Buck-O", what's this I hear about you and a gang in Redcar pubs on Saturday!'

'Who, me, sir?' Doug flustered at the sudden attack.

'I've heard about your drinking and carrying on, young-un!'

'No, sir – not me, sir.'

The card game stopped. Amusement registered on the tired faces at the lad's discomfort.

'Stay out of Redcar pubs, d'you hear – in fact never mind Redcar pubs – stay out of Redcar – d'you hear – eh!'

By now the lad had regained some of his wits and his blushes had subsided. He finished with his sweeping and left, only to return minutes later with a damp cloth to wipe the oil cloth on the opposite table. Joe, thinking to make the most of the situation, returned to the attack.

'And where the hell are you off to tonight, young-un?' he thundered at the lad's back bent over the table top.

The young man swung round from his task to face him and with a radiantly disarming smile, replied,

'Sunday school outing, sir!'

For a second there was absolute silence and then Joe's three colleagues collapsed into roars of smoky laughter. Joe's face blackened, he removed the empty pipe from his clenched teeth where it was in peril of being bitten through and, pointing the stem in the luckless lad's direction, was about to deliver a severe wigging when there was a shout from above and a jolting crash. The cutter had emerged from the fog to land heavily alongside the wooden jetty.

Cards, money, bodies, bottles, mugs, ashtrays flew in all directions, landing in complete chaos on the floor of the saloon.

Ephy rubbed his elbow and yelled to the unseen skipper in the wheelhouse above, 'I think we're there, skip!'

Doug, ever resourceful, had used the confusion to skip away unseen while the card players sorted out their hands, each accusing the other of switching cards.

'I was going a "bundle",' declared John.

'I had a sure fire misère,' lied Ephy, who was determined that John wouldn't walk off with the pot.

'If we don't hurry up we'll be cutting "Ace high" for it,' declared Joe, who was secretly glad of the diversion to save a bit of face. 'If I can get one good hand now I'll play it in the bus going up the road. That kitty's too big to go on the turn of a card.'

The fog gave way to a brightening mist. The disc of the morning sun could be seen trying to burn through the thinning scuds of vapour. By mid-morning it would all have dissolved and vanished as if by magic.

As they strolled up the rickety wooden jetty, Ephy Dee caught the eye of young Douglas. In a sudden pang of paternalistic interest he called the young boatman to one side.

'If I were you, kidda, I'd make my peace with Joe Braithwaite right now before it goes too far. He has a memory like an elephant and could well be still on the Board when you come up for a Licence. He could make things very hard for you. Besides, he's feeling good right now, he's just won that kitty with a lousy seven of diamonds!'

CHAPTER 4

Every day was a working day for us except Sunday. Most of our days were spent in keeping the ship well maintained; by that I mean knocking off rust with chipping hammers or washing paintwork with soap and fresh water. If the weather was bad we worked inside, scrubbing floors, polishing varnishwork with vinegar and water, cleaning brass, changing bed linen and a hundred other jobs that could be thought up by the fiendishly inventive Captain Jack. I sometimes thought if the saloon had had a piano he would have had us tuning it!

But Sunday was ours. Unless we were busy with shipping, of course. But then we didn't mind that so much; it gave us a chance to get off on our own. So, after the breakfast pots were washed and stowed away, it was each to his own whim. Sometimes, if cold and wet, it was a lie down read; sometimes it was a yarn in the engine room and sometimes it was a hunt round for a line for a spot of fishing.

Anchored at the Tees Fairway was a great spot for fishing and often the bites were plentiful enough to keep us entertained. The seas were well stocked just after the war as they had been a battle ground for at least four or five years. Dangers from mines and marauding aircraft had kept fishing to a minimum and anyway most inshore fishermen had been on war service of one kind or another and so the fish stocks in the North Sea had built up and the pickings were rich. It was not unusual for Pilots to stay up all night to fish and go home in the morning with several stone of prime cod or haddock.

Our interest was more of a dilettante nature. The sport and sometimes the opportunity for mischief or horse-play appealed more. The competition was brisk for somewhere to put a line over the side, the trick being to make your line fast to a stanchion or rail so it wasn't necessary to fish it all the time. It was these unattended lines which were the object of most of the pranks.

A favourite trick was to tie the lines of two Pilots who were fishing on opposite sides of the cutter, out of sight of each other. For this you had to wait until someone brewed a pot of tea and the unsuspecting pair left their lines unattended to get their pint pots. Then you carefully lifted the two lines simultaneously, carried them right aft, tied them together and dropped them over the stern of the cutter to let them sink to the bottom again, and hid to watch the pantomime.

Out on deck came the two protagonists, unsuspecting, tea in hand. They placed their pints on the gunnel while they lit a fag, casually lifting their lines almost out of habit, but in doing so felt a tightening tug as each line came fast around the cutter's barnacled keel. Each gave another pull, testing the weight on the line again. Of course, as each pulled the other's line the tugging came stronger and stronger until the cry came to 'Fetch the gaff hook. I've got a big un!'

Sometimes this went on for several minutes before the penny dropped. Then it paid to be going ashore in the motor boat or to be locked in the toilet.

Mackerel time was all action. In the summer, usually July or August, the shoals of mackerel arrived off our coast chasing the even larger shoals of sprats. When this happened and shipping was slack, we often asked permission to put the pulling boat in the water and, with two at the oars and two fishing with spinners in the stern sheets, we would row amongst the billions of leaping sprats. Often the water literally boiled with the little fish in their frantic efforts to get out of the way of the voracious mackerel. Great areas of water would be covered with the leaping, struggling little fish and the boat would be surrounded by a living carpet of fish several feet deep. It is often said, with some justification, that you can't catch mackerel, they give themselves up – almost leaping into the boat. When you are amongst them you pull them in three, four, five, six at a time, depending on how many spinners or feathers you have on your line. Often you don't even need hooks, just a piece of coloured bunting or rag – something for them to catch a hold of while you yank them into the boat. They lie flapping and quivering in great shimmering heaps as you hurl your line back over the side as fast as you can. Pretty soon you are standing ankle or even knee deep in a jittering pile of fish, and plastered in blood and gore as you tear them off the hook. You finish up with far more than you can possibly eat and so we would fill an old sand bag or carrier with forty or fifty good fish and drop them off at friends' or neighbours' houses on the way home. A fresh mackerel makes an excellent meal but they must be eaten fresh. Ours always were.

A completely different kind of fishing but one that, given the circumstances, could be very effective, was sometimes indulged in if conditions were just right and opportunity presented itself.

Occasionally, on calm, hazy days, we would hear the distinctive exhaust note of Danish fishing boats. These small, wooden, stoutly-built craft usually had a crew of three or four and would fish the North Sea for several weeks, navigating with the most rudimentary aids, until they had a good saleable catch and then they would make for the most convenient port with a fish market. For some reason they seemed to favour Hartlepool rather than Whitby, which is some twenty miles to the south-east and has a sizeable fishing fleet of its own and thus the competition maybe that bit keener. Anyway, these little craft would appear out of the mist, often two or three in a group, their ancient single cylinder, hot bulb engine banging away furiously as they nipped along at a spanking eight or nine knots.

On spotting the cutter lying at anchor, they invariably throttled back to a ridiculously slow tickover. The flywheel on the old diesel was so massive and stored so much energy that the effect was almost BANG – two, three, four – BANG – two, three, four – BANG -two, three, four – and so on.

Their leader usually came close alongside, less than thirty or forty feet, and a whiskered face would bellow from the wheelhouse window, 'Vere Hartlepool?'

This was where a well rehearsed charade would be performed aboard the cutter. At first the people on deck, be they Pilots, boathands or foreman, would grin and wave cheerily but completely ignore the question. This would put the Dane in some doubt, firstly as to his command of the English language, secondly as to his position, of which he was already uncertain, despite the fact that we had TEES in letters four feet high along our sides, and, finally, whether or not we were all idiots. His head would disappear back into his tiny wheelhouse whilst he consulted his North Sea general chart, albeit coffee-stained, and filthy with meaningless scribblings, radio frequencies and other irrelevant miscellanea. One such boat boarded was actually navigating on a school atlas!

Muttering and chuntering to himself various Danish fishing obscenities, he would emerge again – 'Vere Hartlepool?' he would repeat.

This time he was so close that he almost got a load of ashes from our stoke hold chucked on his deck by Steve the fireman. Cursing and swinging the wheel to avoid colliding with our motor boat moored alongside he was just about to make a third circuit when Jack Nixon, the senior Pilot, would appear, his face beaming with sudden enlightenment – 'AHHHH, Yes – Hartlepool, I come on board, Captain.'

This was the signal, and what followed was almost akin to piracy. All three boathands leapt into the boat followed by a couple of Pilots. The engine was cranked into life and off we went alongside the bewildered fisherman, scrambling on board as best we could.

Jack Nixon and his partner, a much more junior man, smoothed themselves down to regain as much dignity as possible after the hasty scramble to reach his deck. Everywhere was littered with coils of rope, drums of lube oil, wicker baskets, rusty tools and everything plastered in fish guts and dried-on scales. The two men squeezed into the tiny wheelhouse like policemen in a telephone booth on a wet night.

'Well now, Captain,' Jack flannelled the grizzled skipper in a manner usually reserved for Council Chamber Meetings, 'Do you have any nice fish for us?' The question was quite academic as the battered hatch was whipped off to reveal a hold full of prime fish.

'Ya, ya, ve have a little cod.'

The biggest basket we could lay our hands on went down into the stinking hold, closely followed by the junior apprentice, who proceeded to pull out the finest looking fish he could lay his hands on and fill the basket to overflowing. The two Pilots kept an eye on the proceedings while occupying the skipper in a high level conversation on his navigational deficiencies, real or imaginary. A diagram of Hartlepool entrance was drawn on the back of a cigarette packet and he was given a course to steer. The younger of the two Pilots inspected with ill-concealed derision the minute, battered compass, carelessly slung in home-made gimbals in the corner. The compass card with its points and quarter points was so faded as to be almost illegible. As a navigational instrument it was little better than that used by his Viking ancestors in these same waters a thousand years before, only they didn't come to sell fish!

So, after much hand shaking and back slapping, the Dane, well satisfied with the advice given, helped the elderly Pilot back into the boat which was by now covered in slithery fish scales, the contents of the basket having been tipped unceremoniously into the fore part. The whole boat was littered with several stone of big cod and the mess was considerable.

As we departed from his side the Nordic fisherman cupped his hands and bellowed across the water to his companions lying stopped a few yards away. Waves and shouts followed and, together, they opened up their throttles, the big Bolinger diesel accelerated to a rhythmic BOP-BOP-BOP full speed, in the course of which it blew several perfect smoke rings into the still air from its vertical six inch exhaust pipe. In unison they swung round onto their new course to be confronted by the gleaming white lighthouse of Hartlepool Heugh, shining brilliantly in the sunshine a couple of miles away. The fog had lifted with all the suddenness of a roller blind and in a matter of seconds all was revealed as the sea and the sky drew a straight edge at the horizon.

Old Jack, his faded mac covered in fish scales, grinned as we sped back to the cutter.

'He shouldn't have much trouble finding Hartlepool now,' he quipped, 'If he doesn't look out, he'll run slap into it. Save me three big uns, son,' indicating the fish he had his eye on, 'and don't forget to wash the boat out when you've finished.'

'Aye, aye, Mr Nixon,' we chorused.

Jack was well known for his impartiality when sharing equal shares. After all he was the originator of the expression, 'They're all the same but that one's mine.'

CHAPTER 5

Very early on in our boathand days the need to use the Aldis lamp efficiently was impressed upon us. It was our sole form of two-way communication with either the shore signal station, or perhaps more importantly, ship to ship.

Now an Aldis lamp, for those who don't know, is like a miniature hand-held search light. It is fitted with a pistol grip and a trigger which is coupled directly to a concave mirror immediately behind the powerful bulb. On gripping the switch incorporated in the handle, a parallel beam of light is projected from the business end. Signalling, in morse code of course, is achieved by bringing that beam of light to bear on your target viewed through a sight arranged on top of the lamp itself. Its great versatility lies in the fact that it can be used equally effectively in daylight and darkness so long as it is aimed properly. The beam does not shut off and on like, for instance, the much weaker, all round morse signalling light, fitted on some ships; it is directed on and off its target by the trigger tilting the mirror. The whole thing is quite small, about the size of a small saucepan.

Until reaching an acceptable standard of signalling, we spent many hours on the deck of the cutter, signalling and practising procedures with our boat torches.

We soon became expert and often needed to be, as signalling morse left a lot to be desired with some of our more Mediterranean visitors. Not so with the keepers of the light at the Gare. They were expert, but surprisingly enough, sufficiently individual in their technique to be recognised as being Ike or Charlie working together on one watch or Little Johnny or dour Jack on the other watch. We could nearly always tell who was on the other lamp. These keepers, there were only four of them, worked the whole time two days on and two days off the same as our skippers, engineers and firemen.

As we manned our Aldis lamp, the messages would be flashed between us and the shore at such length that our eyes would stream with tears at the strain and concentration needed not to miss a single dot. We took a great pride in receiving each word first time and not asking them to repeat. Lengthy messages required someone standing by to take down on paper as the reader deciphered each separate word, otherwise, having successfully reached the end of the message, you forgot what the beginning was about.

We never had this problem with shipping. Our only concern with calling up ships was to find out 'What Ship?' and often 'Where Bound?' We seldom entered into lengthy conversations as the signaller on the bridge was often on his own and had better things to do than flash away at us and anyway, often as not, he only knew his own ship's name in morse! To pursue the matter further was a sheer waste of time. Vessels of the British Tanker Company were one obvious exception. They placed great importance on good signalling and practised it at every opportunity. We were not a tanker port at that time and so our only contact with them was as ships that passed in the night as they steamed south, miles offside, on track from the 'Tyne to Mina Al Ahmahdi' or 'Grangemouth to Ras Tannura'. At the time their destinations had an Arabian Nights romantic ring to them.

Often ships we called up sent unreadable replies and a great deal of imagination or knowledge of what orders we actually had in the arrivals drawer was necessary to translate and decipher the pyrotechnics we were subjected to.

Sometimes we failed.

One ship I remember vividly had everyone on the cutter flummoxed.

Peter, on watch until about midnight, spotted a pair of dim mastheadlights approaching from the east. It was one of those velvety black nights with no moon and only an occasional glimpse of the stars as they peeped shyly through the high stratus cloud layer.

Peter groped and fumbled around in the dark wheelhouse and eventually found our ancient brass night glasses. Giving them a cursory rub on his jumper he tried to focus them and eventually just managed to make out a red side light. They were clearly the kind of navigation lights that you had to strike a match to see if they were on!

He reached for the Aldis resting secure in its clip high above the wheelhouse window. Sliding back the door, he rested the lamp in the crook of his arm and began signalling.

Down below, in the fo'castle, we who were trying to sleep, heard the familiar clackety clack of the Aldis mirror as the slim pencil beam of light pierced the darkness.

Several minutes elapsed. Peter persisted. We could hear him muttering to himself as he pounded away. The ship's lights were so poor that he had to check constantly that he had his lamp focused in the right direction as the scattered light from the powerful bulb tended to blind him while he was signalling. Clackety clack, clackety clack, he paused again. Had he seen a flicker of light from the direction of the ship's sidelight? Maybe not. Staring into the darkness sometimes plays strange tricks; you can imagine all sorts of things. Weird forms and shapes dance in the darkness of your eye.

Suddenly a tremendous shaft of light transfixed the cutter and bathed the whole of the little craft in a brilliant searchlight beam. There followed a barrage of morse, more or less directed at us, from what must have been an enormous signalling lamp. Every few seconds the light would gyrate wildly about the night sky as the signaller seemed to lose control, regain it, refocus, then blast away again.

Peter yelped in surprise at this overwhelming onslaught. Clearly the stranger was equipped with a superior weapon! His cry brought us up on deck to see what was going on. Peter stuck to his guns. Please repeat, dit dit dah dah dit dit. And repeat he did, this time with even greater gusto – Flash – Flash – Flash, utterly unreadable morse. We shielded our eyes against the terrific glare.

By this time Pilots began appearing on deck, wondering what was going on. Some tried reading the blinding light and offered the odd syllable in advice to the confused and angry Peter.

'INGO!' shouted Hedley.

'PAT!' shouted Sam.

'KATI!' shouted someone else.

The searchlight flashed away furiously. By now it seemed there was no way of stopping it. We were being assaulted by a phantom signaller. Maybe an updated version of the Flying Dutchman? The vessel steamed closer and closer, all the time signalling furiously at us. We seemed to have a tiger by the tail!

Then suddenly all went black again. The searchlight was extinguished. He was by now so close we could see the massive filament in its bulb glow red before it finally died.

We all blinked in the enveloping gloom and eventually made out the shape of a large ship less than a quarter of a mile from us. As we watched, a faint glow of light blinked from a position over his bridge. It was his puny morse light mounted on a short mast above the wheelhouse. The lamp spelt out the single letter G – dah dah dit, the international signal for 'I require a Pilot'.

'Stand by the boat.'

Apple Sam, the Pilot on turn, plunged into his oilskins as Peter abandoned the Aldis and skipped over the boat deck to man the steam winch.

Arnold and I dragged on our sea boots and, grabbing an oilskin each, slacked off on the relieving tackles that bowsed the boat in tight to the rail.

'Lower away,' we shouted in unison to Pete on the winch.

There was a hiss of steam, a couple of hefty bangs as the steam contraption cleared its throat and then, with a clattering roar, the boat dropped neatly into the water.

We leapt in, yanking the release gear to free the fall wires. The tiller was slammed into its brass socket and the engine cranked into life as we cast off the bow line and swung away from the cutter's side, all in a continuous, logical sequence that practice had made almost automatic. We developed a kind of deft opportunism when in and around our boats.

By always rolling with the motion, leaping in just as it is beginning to move down, and in other ways developing a sense of timing and rhythm, it becomes second nature and makes boatwork look much easier than it really is. People not used to this and not having this sixth sense, by comparison looked clumsy and awkward. It comes with continuous practice.

The point was made years later, when another boathand and I were away at sea together as cadets in a Liverpool Liner Company. We were taking some passengers for a picnic in the motor lifeboat and Alan and I were detailed off to go with the 3rd and 4th Mates. Both these gentlemen were younger than we were but, being big ship men, had had no small boat experience, and it was very obvious. Peggy, the Captain's wife, remarked afterwards that we were walking about like a pair of cats whilst our superiors had to sit down most of the time. We had a quiet smile to ourselves.

We buttoned up our waterproofs as the boat ran easily across the black water, closing the distance rapidly between us and the darkened ship ahead of us.

As we approached, Sam said to my companion, 'What's your torch like, Arny?'

'It's a brand new one, Mr Wilkinson. Why, what do you reckon it is?'

'Well, Hedley reckoned it's the *Inga*, but we haven't it due. Might be that old Greek, the *Kate C*. He seems to like to sneak up on us, but I don't think he has a searchlight like that, 'course, he might have just bought one and can't use it yet! Anyway we'll soon see. Give us your torch here.'

Arnold handed his precious new flashlight to Apple Sam as we ran close under the towering sweep of the ship's bow like David and Goliath, and shone the Woolworth's torch up over our heads.

Right at the limit of the light's effectiveness, we could just make out a name, and unconsciously chanted in chorus,

'KATINGO HADJEPETARAS'.

'Of course,' chuckled Sam, 'I might have known.'

'Bloody hell,' giggled Arnold, 'Wait 'til we tell Pete!'

CHAPTER 6

The need for a torch at night in an open boat was absolutely vital: much more so if the weather was bad and we had to resort to a pulling boat in order to board ships safely.

It was the lad left on the cutter who had to man the winch and, while the others were away in the pulling boat, had to keep a lookout and keep the cutter's skipper constantly informed of our position as we flashed our light every few minutes.

We strived to keep our flashlights in working order and there was a constant searching around for one that worked. Corrosion, accidental damage, rough usage and general wear and tear took its toll and sometimes we were hard pressed to find one that worked properly.

The Pilot Office issued them, and were loathe to spend money on decent waterproof, rubber ones as the mortality rate amongst them was high, so they issued us with cheap, metal things that didn't last. I used to cut a length of bicycle inner tube and pull it over the torch like a sleeve in an effort to preserve mine.

When boarding ships on a pitch black night, we needed to give the ship an indication of our presence in order for him to present a good lee for us. It was necessary for the ship to be almost stopped for us to row along his side to pick up the boat rope. This technique was long and drawn out and nobody enjoyed it as it involved a lot of hard work, careful positioning, precise seamanship and quite an element of risk.

So we tended to use the motor boat for as long as possible in bad weather, huddled behind a staunch canvas hood, squinting against the stinging rain squalls and dodging the sea water as it came whipping silently over the bow to lash back at us.

A bunch of characters on the boatdeck

Cutter Confusion

One black night the cutter lay at anchor behind the shelter of the Breakwater; a nasty easterly sea was running outside. Seas were pounding the lighthouse with the occasional flurry of spume and spray splattering the lantern itself as its beam circled the horizon. About two in the morning the lights of a ship were seen approaching the Fairway Buoy from the north. As the weather showed no immediate signs of improving it was decided to pick up the motor boat and get the cutter underway to go off and board with the pulling boat.

The use of a pulling boat was necessary in bad weather because it was much, much lighter than the motor boat, being, of course, minus an engine. Any attempt to drop the heavier motor boat or, even more important, to pick it up in a bad sea, could have resulted in the boat being 'bounced' in the falls. Any sudden vicious jerk could cause the heavy engine either to break the boat's back or, in the extreme, to go straight through the bottom of the boat. Heavy metal springs on the end of each fall, which were designed to cushion this kind of thing, could only take so much abuse before they became a hazard and in danger of breaking.

On one occasion, when using the pulling boat, the falls and springs took such a severe hammering that one spring shattered and flew in all directions, a piece weighing four or five pounds passing within an inch of a Pilot's nose!

Everybody turned out to lend a hand when the cutter was boarding in bad weather. The younger men helped with the winding in and out of davits and the older ones usually stood about offering advice or lending a hand where needed. There was no comfort to be had anywhere on board so they might as well be up and about watching what was going on.

The motor boat was hoisted as high as possible to the very heads of the davits and then, with great effort on the handles of the patent winding gear, was wound inboard and landed in its chocks on the boat deck. Clamps over the gunnel to senhouse slips attached to lugs in the deck made sure it was well secured and couldn't move. Meanwhile the winding gear on the steam winch, also on the boat deck, was changed over to facilitate launching the pulling boat which sat on its chocks on the opposite side to the motor boat.

One of us would climb into the pulling boat to make sure that everything in it was secure and to hand. Oars were lashed and yet handy for instant use, rowlocks were unshipped but ready, bailer handy and plug in (very important!) and, last but not least, the long painter was well fast through the snotter and round the thwart with the customary round turn and two half hitches.

We once lowered the pulling boat in a terrible panic only to find, when the falls were released, that the boat was not made fast properly to the cutter's boat rope and drifted away with water pouring in through an open drain hole. The poor soul who had jumped in to pull the handle of the simultaneous release gear, letting go the falls, had to scull it back with one oar, knee-deep in icy cold water.

As the cutter plunged out from behind the shelter of the lighthouse, the senior lad went up into the crowded wheelhouse to contact the arriving ship on the signal lamp. It was necessary to warn the Master that we intended trying to board him with a pulling boat so that he could make certain specific preparations.

These preparations consisted mainly of having a guest warp or long boat rope from right for'ard through a lead at the break of the fo'castle on the lee side. This must lie in a great sweeping curve or bight along the ship's side and the other end made fast well aft. The pilot ladder must be rigged over the same lee side at a point where the boat rope almost touches the water. Having rigged this gear, together with lifebuoy and heaving line in case of accidents, the Master must now so place his ship with the sea on the opposite quarter to the side the ladder is rigged and have minimal but slight head way. (It's a hell of a job trying to row a heavy boat as fast as a ship's slowest speed downwind!)

As it happened, this night, the ship was a regular visitor, Swedish and an excellent seaman well versed in these techniques. Very often we were not so fortunate.

The next part of the saga concerns the Pilot cutter and its crew. The cutter now has to arrive at a point sufficiently upwind of the ship to be boarded, so that the boat, once dropped, can more easily travel downwind. He has to make a good enough lee to drop his boat and its crew safely, then travel downwind himself, clear of the ship, to be able to pick his boat up again after the Pilot has been boarded.

Sometimes, if the vessel to be boarded was big enough and made a good lee, the cutter's skipper would steam close up under the ship's lee but at right angles to her and heading directly away. This manoeuvre enabled him to keep a good eye on his lads in the boat who had to row the shortest possible distance to be picked up and finally, once picked up, the cutter was in an ideal position to steam clear of his 'customer' who would most likely be stopped in the water and blowing down on him very fast indeed!

This may all sound rather complicated but, in reality, it was a matter of common sense, good seamanship and experience. It was often done at night, frequently in rain or snow and sometimes bungled! It was to the credit of all concerned that accidents were extremely rare, although we had some very hairy moments.

By now it was obvious that the weather was rapidly deteriorating. The wind was blowing a full gale and the sea had lost its boisterous bounce and was becoming nasty and vicious. Slowly, the cutter ploughed her way out, burying herself in the steepening seas and occasionally shipping one right over the stem. The water came cascading and crashing over the anchor windlass and then accelerated down the narrow decking as the vessel lifted on the next sea, tilting the deck to crazy angles, finally sluicing over the sides through the open door in the bulwarks.

As was usual in these situations, the Pilot on turn for the ship we were going to board was with the skipper in the wheelhouse together with several of his colleagues. The debate going on was whether to attempt to board in the dark or to wait until daylight (things always looked better in daylight). Someone had given the barometer a tap on his way up onto the bridge and noted its sudden fall, indicating that it would be unlikely to be better come morning.

Leonard Bickerstaff, the Pilot on turn, observed, dryly, that as it was such a black night, we couldn't see enough to frighten ourselves, so why not try to board right now.

'The night's buggered anyway!' he remarked succinctly.

To their credit, Pilots would often ask the boathands how we felt about the weather. After all, the Pilot only had to do a one way journey. Once aboard his ship he was secure. We had the return journey to do with the additional hazard of picking the boat up and securing it.

We huddled in the warmth of the galley, the stable door closed, as the decks outside ran awash, ankle deep with successive seas being shipped. Salt water hissed and slapped against the gunnel and bulkhead as the cutter crashed and wallowed in the seaway. In the wheelhouse, Pilots and skipper alike peered out of the wheelhouse windows into the blackness. It was raining now and the rain ran down the windows in little rivulets; every so often a sudden gust of wind displaced them all like an old movie film jumping a couple of frames.

Eventually we arrived at our rendezvous with the Swedish steamer *Radmanso*. We heard the rattle of the telegraph wires as they clattered over our heads. The steering gear rumbled high on the bulkhead as Jack Emmerson began to place his charge in a good position to drop the boat.

'Stand by the boat!' came the shout.

We scampered along amidships, up a vertical, iron ladder and onto the rolling, heaving, boatdeck. At a nod from the skipper, looking aft through the wheelhouse window, we each knew our station: Pete on the steam winch; Arnold and a Pilot on the forward davit handle; myself and a Pilot on the aftmost one. The boat was hoisted out of its chocks and, dangling by its fall wire, started swinging wildly. No time to do anything now except wind like fury on those handles to swing the boat over the side, for this is the time the boat can damage itself with no relieving tackles to restrain it.

The grim face of concentration peered out at us from the wheelhouse as the skipper tried to assess the precise moment. Timing was the essence of success.

Leonard Bickerstaff, or 'Bick' as he was called, his tram driver's hat screwed on tight, overcoat buttoned up, stood resolutely awaiting events to unfold. The davits were swung out to their extreme. The moment was near. We all braced ourselves against the gale and the rolling deck.

Cutter Confusion 51

'Lower away!' screamed Jack.

Peter slammed over the massive iron lever on the winch and turned on three full turns of the steam valve at a single movement. The winch clattered into motion, steam belching from its blurred piston rods as it accelerated under the wide open steam valve.

Arnold, with the agility of a cat, stepped into the fast descending boat as it passed boat deck level, stuck the plug in, whipped the outboard rowlock into its housing, the lashings off the oars and grabbed the release handle.

Crash! The boat hit the water and instantly the falls were released. As the head rope came tight with a jerk, I leapt into the fore end closely followed, some seconds later, by old 'Bick'.

'OK, son,' he muttered. 'Cast off and bear away,' taking the weather oar himself while we got ourselves sorted out.

The night was pitch black and the boat danced about like a cork as we pulled steadily towards *Radmanso*. Seas came dashing at us out of the darkness, breaking and bubbling, threatening to engulf us at every rush but skidding harmlessly under our bobbing, high bulwarks as the boat's inherent buoyancy overcame the sea's clumsy, tumbling weight. A quietness descended on us after the uproar of the launching. There was the creak of oar in rowlock, the breathing, heavy with exertion, and the slop and gurgle of water under the boat's planking.

Leonard broke the silence, standing in the stern sheets where he had positioned himself well clear of the rowers.

'Shine a light, son.'

'Give 'im a flash,' Arnold turned to me.

'The bluddy torch's in the bottom of the boat somewhere, rolling around. It's smashed I think.'

'Christ – Pete's got mine,' muttered Arnold.

'Shine your light, son, or he'll never see us.' 'Bick' spoke again in his curious way out of the corner of his mouth.

'The torch is smashed, I think, Mr Bickerstaff,' I gasped.

'Hell, son, he'll run us down if we don't show him a light – where is it?' He grovelled round in the water in the bottom of the boat, found the

torch and emptied about a cupful of water out of it through the smashed glass.

'Why don't they get you decent torches instead of this rubbish,' he muttered half to us and half to himself, and then, 'I'll strike a match, son – it's black enough for 'im to see that.'

Some hope, I thought, in this wind.

But remarkably enough I was suddenly conscious of flames behind me and, looking round, was amazed to see the old Pilot setting fire to crumpled pieces of paper he was producing from his coat pocket and throwing them in the air before the flames burnt his fingers.

Arnold and I looked at each other in amusement as 'Bick' persisted with matches and bundles of paper he kept producing like a conjurer produces doves.

'Keep rowing, son, he's seen us – OUCH!' 'Bick' burnt his hand on another bundle of flaming paper as he threw it in the air astern of us.

What the Swedish skipper must have thought at this display was anybody's guess. He must have wondered what was going on.

All of a sudden we were alongside the black steel side of the ship. Arnold unshipped his for'ard oar as I pulled the bow of the boat hard up against the ship's side with my oar. As we came level he grabbed the boat rope, hanging conveniently, and took a turn under the hook and round the post in the boat's bow. We now had the boat under control alongside and it only remained for the rope to be slacked gently until we were alongside the ladder. I kept the boat hard alongside with my oar while Leonard clambered over the thwarts to reach the ladder.

'Are you OK, Mr Bickerstaff?' Arnold said as he laid hold of the white painted rungs to steady himself.

'Yes thanks, lads,' 'Bick' said out of the corner of his mouth, 'But I don't know what the "old man" is going to say when I get on the bridge – I've just set fire to his mail!'

CHAPTER 7

Alongside the black, steel hull of an ancient steamer is not the most comforting of places to be in the middle of a winter's night.

To two oilskinned lads, bathed in the pale light of a cluster of bare bulbs, struggling with every nerve and fibre to overcome the silent forces that pinned them there, this was an invisible trap.

The only way out lay in their own hands, it would need all their enthusiasm, strength and boatmanship.

Although we were aware of unseen eyes watching us from behind the glare of those bulbs, nevertheless it seemed an awful, lonely place and the sea in the blackness beyond the ring of light also had a hidden menace to it.

The immediate problem was often one of actually getting clear of the ship's side. A loaded ship, deep in the water, was a lot less trouble than a ship in ballast, or 'light', to use the common expression. In the process of making a good lee for the boat, the ship had to place wind and sea on the opposite side and, in consequence, made a lot of 'leeway' or sideways movement downwind. If the ship was light, this leeway could be considerable and made it almost impossible for a small boat to get away from its side as it blew down on top of it.

We were left in the boat to sheer off as best we could. The technique involved the man for'ard slacking off the boatrope from round the post and hook and pulling the boat ahead on it as fast as he could with all the strength he could muster before letting it go to fall back in the water. Whoever was aft would have his lee oar in the water in such a way that by laying back on it with all his might, the boat would veer off the ship's side with this initial and vital momentum. Team work and timing came with practice.

Once clear of the ship's side, we had to pull like maniacs to maintain our slight advantage as the Pilot, hopefully by now on the ship's bridge, dared not move its engines until the boat was well clear. If the Captain on the bridge did not realise our predicament and set the engines 'Full Speed Ahead' before the Pilot arrived, out of breath, on the bridge, to take charge, then we were in trouble.

If the leeway made by the ship was too great for us to overcome, the only solution was to try and bring the ship around head to wind. But, of course, this was easier said than done with a low powered steamer, and, once we had let go of the boatrope, we, at least in the boat, were committed.

A ship in ballast invariably had almost half its propeller out of the water. This put us in the position of poor little Vera and the great big saw coming nearer and nearer. We slid further and further aft down the ship's side towards several tons of wildly thrashing propeller blades threatening to make a good job of us and the boat. This was always a very adrenalin-stirring exercise and one guaranteed to make us bend those heavy, ash oars!

So we strained ourselves to the utmost as we cleared the circle of light down the ship's side. Once we were clear, the cluster was taken inboard or switched off. We were alone again, rowing in the darkness, and looking towards our parent cutter that was backing up towards us, shortening the rowing distance as much as possible, as it smashed stern first into the confused seas, sending such great clouds of salt spray high into the air that it almost disappeared from view.

These nights could be bitterly cold and although rowing made us sweat inside our oilskins, our hands, if unprotected, could quickly become blue and numb. One little trick learnt from the veterans amongst the Pilots was to bring a pair of old, woollen socks in the boat. We never ever wore gloves when working as it could be very dangerous when tending running wires or making fast. Many a man has lost a finger or, indeed, his hand, through a glove becoming caught in a fast moving wire or crushed by a mooring rope on the drum end of a winch.

Towing was better than rowing

But, when rowing in bitterly cold weather, the cuff of a frayed oilskin rubbed and chaffed the backs of sore hands and could be extremely painful. We dunked the old, woolly socks in the sea and then pulled them over our hands and in a few minutes they were as warm as toast.

We rowed steadily and rhythmically. The sea slopped and hissed around us. We braced ourselves, feet wide apart on the bottom boards. The bilges ran with several inches of water, splashing above the boards, almost ankle deep. We had to watch carefully each successive stroke to make sure our blade went into solid water, so violent was the boat's motion.

Great seas reared up astern of us and, like a giant taking deep breaths, they swelled and swelled until, unable to contain themselves, they burst in great, tumbling flurries of cascading foam. Our cockleshell lifted and skidded away from them. We pulled away, silent, except for the grunts of exertion. The distance to the cutter seemed endless but, slowly, the gap closed.

As he spotted the returning boat, the skipper turned his craft to offer to us as much relatively calm water as possible. He then had to decide the best course of action to take. He could elect to try and pick it up and risk damage, or he could tow it back into shelter before picking it up.

As a rule we only towed the boat as a last resort as we ran the risk of losing it if the tow line parted and, also, the cutter was handicapped if any alterations of plans had to be made.

We arrived back alongside, tired and sweating. The sea was still confused and we bided our time as large lumps of breaking seas threatened to dash us against the steel bulwarks. The cutter wallowed and rolled as it sluggishly answered the helm. We rode the crest of a big wave almost at boat deck level as we swept forward with a tremendous rush.

Helping hands on board threw us the boatrope. The skipper appeared on the wing of the bridge, jackbooted, holding his battered cap on against the wind.

'Look smart, lads,' he bellowed against the wind. 'We'll be out of his lee in a minute.' He pointed to *Radmanso* beginning to gather speed a few hundred feet away.

Cutter Confusion

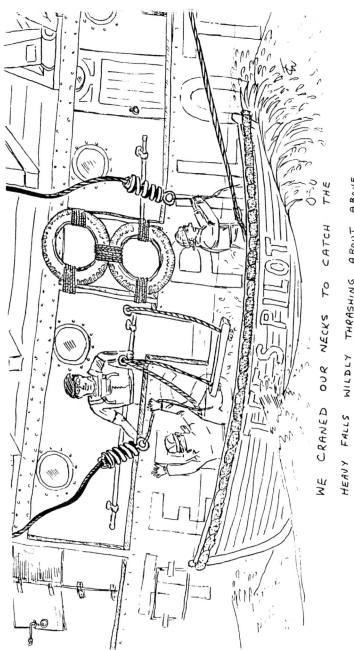

We craned our necks to catch the heavy falls wildly thrashing about above our heads.

The steam winch thundered away, sending the heavy fall wires down towards us in each end of the boat. We craned our necks to catch the heavy spring, with its retaining ring and swivel, wildly thrashing about above our heads. A clean catch and a quick connection with the counter-weighted hook at each end of the boat was something which came with practice. A badly fumbled job would often reward us with pinched and bruised fingers or hands torn on snags sticking out of the serving on the wire falls.

The boat took a vicious run down the cutter's side on another swell as we leapt clear through the bulkhead door and back on board the cutter.

The boat was now in the hands of the winch man. He must judge the roll of the cutter just right, weigh the boat as gently as possible in the falls and then whip it out of the water as quickly as he could before another wave had a chance to crash into it from below. Once out of the water, the boat was hoisted to the limit of the falls, right up to the davit head and then wound in as fast as possible until we were able to steady it and land it once again in its retaining chocks, finally clamping and lashing it to the deck.

Everyone heaved a sigh of relief as the skipper popped back into the wheelhouse, switched the boat deck floodlight off and prepared to run back inside. The older Pilots retired to the comfort of the saloon and we boathands made a bee-line for the warmth of the galley.

Now, once a ship turns round and runs before the sea, as opposed to ploughing into it, it presents a whole new set of problems and dangers. It is not enough to point your nose in the direction you want to go and press on regardless, a following sea can be a treacherous adversary. In our case the two main dangers were either broaching to, when a following sea swings you round and you are in danger of being rolled over on your beam ends, or pooping seas, where heavy, breaking water rolls over your stern and then breaks on the after deck, depositing tons of wild water on top of you.

Captain Jack's immediate concern was controlling the vessel's speed so that it didn't run off its helm and take a sheer one way or the other. Careful helmsmanship was needed, anticipating rather than correcting,

watching the run of the sea as it piled up astern and presenting the stern as square on to it as possible. As far as pooping seas are concerned, well, there's not much you can do about that. If a sea decides to break over your deck, then it will, and nothing you can do will stop it.

This was, in fact, our Achilles heel. We didn't have enough speed or a quickness of helm to outrun the biggest seas.

We pooped two in quick succession. The second one broke over our after deck before the cutter had time to lift herself and get rid of the first one. The second sea was the biggest and did the most damage. It poured down the ventilators into the engineer's quarters. It poured down vents into the engine room and stokehold, washing all the coal out of the side bunkers. There was nearly two feet of water and coal swilling around in the stokehold, and we were lucky it didn't put the furnace fires out. The engine room had over a foot of water in it and the bilges were filled with coal and dust.

Luckily, the engineer was able to keep her going and we limped in past the lighthouse with our stern very low in the water. It was decided to proceed much further up river to the iron ore berth at Redcar Wharf to pump out and have a good look around.

Little Geoff Woodcock, the fireman, got the fright of his life as the water and coal came crashing into the stokehold. Thinking she was foundering, he leapt up the fiddly ladder and never stopped until he reached the wheelhouse. Hurling the door back he appeared wild eyed and covered in coal.

'We're sinkin'! We're sinkin'!' he gasped at Captain Jack.

Jack Emmerson gave him a cold, hard look and then, in a matter of fact voice, replied, 'Your end might be, but mine's OK!'

CHAPTER 8

Occasionally, when the weather deteriorated very quickly without much warning, we could be caught at sea in the motor boat in a situation where we could not pick it up.

One Sunday morning we arrived at the breakwater with the relieving watch of Pilots in the ancient Humber Super Snipe shooting brake run by the taxi firm. This was a remarkable vehicle with enormous balloon tyres of the type used in the desert during the war. We all piled in this car, the watch of Pilots, engineer, skipper, fireman, lads, everybody. We sat in three rows on bench seats. The man occupying the seat next to the driver, who was squashed flat against his door, had to change gear, if he could find the gear lever. As often as not it was up someone's trouser leg!

On the way down I had agreed to give fireman Jack a hair-cut in the stokehold when we had a chance. As we all tumbled out of the old Humber and walked down the wooden jetty, there was a chill easterly breeze blowing with a definite hint of a snow flurry. The sky was as grey as the sea as they merged at the horizon. The cutter lay at anchor between the channel buoys. We clambered down the vertical iron ladder into the M10 or the 'plank' as we nicknamed this particular boat because of its lamentable lack of free-board and its happy, carefree way of getting us wet through. One of the first things anybody who knows anything about small boats will tell you first is – always sit down in a boat. We never or almost never sat down. I suppose it may have something to do with the remark once passed by an old and venerable Pilot who said, 'Wet arses invite piles.'

To the casual observer it must have presented an amusing spectacle,

Cutter Confusion

– nine or ten people all standing bolt upright and being propelled at a rate of knots across the water between jetty and cutter in a boat that was almost level with the water. Sometimes, if we felt extra frisky, we might deliberately start rocking it in unison – until a wave lopped over the gunnel and the senior hand got a shoe full of water.

We arrived alongside with a blast of full astern; there was a mad scramble to vacate the boat and be first in the frying pan with bacon and eggs. Needless to say, by the time we had the boat moored and done all the right things, the galley was crammed with blue reefers and brass buttons. Our place in the pecking order was definitely last, unless we could outsmart the engineer and fireman. This we tried to do, especially the firemen. Their culinary habits were sometimes weird to say the least, and usually meant we had to cauterize the frying pan before we deemed it fit to use again, but, as a rule, we weren't too fussy.

So, Sunday morning, with little shipping activity, adopted a pervading sense of lethargy. The cutter swung at her anchor cable with the occasional clack as the links rattled in the hause pipe; now and again a metallic screech as the furnace door was swung open and the fireman inspected his charges. Not too little, not too much, the water level indicated in the gauge glass, the flicking pointer in the steam pressure gauge not being allowed to rise too high in case she blew off with a thunderous roar, up her steam exhaust, thereby shattering the peace of it all. Snores and coughing from the sleeping berth. All the time the wind freshening.

After washing all the breakfast pots and seeing the galley fire well tended, we were free – until needed. Arnold had a pair of white rubber thigh boots of which he was very proud, and spent a lot of time keeping them clean. Peter sat in the wheelhouse doing the crossword. I went below to give fireman Jack his short back and sides in the warmth of the stokehold. A picture of calm and tranquillity. Meanwhile, the grey waves were getting little white tops, and the cutter was starting to roll just a little, in a long swell that was sneaking round the lighthouse.

It was Peter, at his vantage point in the wheelhouse, who spotted it first.

'Ship at the Fairway Buoy!' He popped his head down the open

ventilator and shouted his message into the saloon below.

Old Jimmy Powell, who was sitting with half closed eyes enjoying his third pipe since breakfast, opened his eyes, removed his pipe and retorted, 'What ship is it, son?'

Peter, having removed his head from the ventilator, did not hear him. With a groan, Old Jim heaved himself to his feet and made for the saloon door. Once out on deck, he walked aft to see for himself the cause of all the commotion. There, standing out starkly black against the greyness of the North Sea was the old Swedish Turret ship *Vindo*, rolling her side decks awash as she steamed into the bay with a gut full of iron ore from Narvik. A splash of yellow on her halyards proclaimed she was 'from across', and as we stood and watched, a 'G' flag as big as a blanket was hauled up to fly alongside the 'Q'. He clearly wanted his Pilot.

'Is the boat full of petrol, son?' The remark from Old Jim was addressed to me as I poked my head out of the fiddly door. Down below fireman Jack, who was just at the pudding basin stage, dozed on an upturned crate.

'There's plenty in the tank and a spare can, Mr Powell,' I replied.

'Get your gear on then, son, and we'll run off to him,' said Jim between pipe-clenched teeth, and, turning on his heel, he went back for'ard to put on his oilskins.

Within minutes we were in the M10, and, with canvas hood raised, ploughing off past the shelter of the lighthouse, into the open sea. It quickly became obvious that conditions were worse than we thought – and deteriorating.

As the boat bashed and rolled in successive head seas, we dodged behind the hood as spray and water hissed past our faces. Soon I was working away at the bilge pump as Peter, at the tiller, looked anxiously down at the level of water rising in the bottom of the boat below the fast spinning propeller shaft coupling. We must keep the level below that item or else we would get very wet from sea water coming up at us from below. I pumped away furiously with one hand, hanging on for grim death with the other. Old Jim, never one for wasting words, uttered the understatement of the year.

'Reckon we're gunna get wet, son.'

Just as he said this, a very wet wave hit him full in the face, his pipe went out with a sharp hiss and from then on he kept his head down.

Now, these so-called Turret ships were a very strange breed. End on, they looked a bit like the neck and shoulders of a bottle, and along their sides about four feet above the water line when they were loaded, they had a long, flat shelf about ten feet wide. This made them very strong, rigid ships and safe in very bad weather, but in bad weather they were devilish difficult to put a Pilot on board. It had been said that the only way to board them when the sea was bad was over the poop, but we didn't think much to that.

The fact remained that here we were in a bad weather situation with a decision to make. Did we try to board in rapidly worsening conditions or did we turn round and run before the sea, back inside and get the steam cutter underway? The dilemma partially resolved itself as I observed the '*B.O.*' underway, belching black smoke, steaming out after us.

By now, the sea was really bad and deep troughs in the swell completely hid the *Vindo* from us. The tops were starting to curl over and break. We had to throttle back at just the right time to give the bows of the boat a chance to lift and clear the breaking white water. If we weren't careful, a breaking wave would fill the boat, and stop the engine, then we would be in real trouble. I released the lashings on the oars – just in case – and at that point everything was blotted out in a fierce snow squall. The wind came away with a dramatic suddenness and the snow and hail stung our faces as it came at us horizontal, brand new off the North Sea.

We all squinted through the murk of the squall to make out the loom of *Vindo*. The ship's Captain, who was having troubles of his own in trying to manoeuvre his low-powered steamer near the Fairway Buoy in atrocious conditions, peered out from the shelter of the small cab built on his bridge wing. Straining to catch a glimpse of the tiny Pilot launch, almost invisible amongst the breaking white water, he heard a cry from the fo'castle head, which made him turn to the direction indicated by the look-out. There was the boat, approaching fine on the port bow; now he must try to make a lee with his heavily rolling ship.

'Keep off her, son.' Jim's command was to Peter at the tiller.

'He'll have to make a lot better lee than that before we dare go near him.' Pearls of wisdom!

The seas ran down his side, deckings well awash, rolling tons of water over himself with each corkscrew motion. The flags on his halyards were coloured boards as they smacked and crackled in the gale. An echelon of gulls kept station with his top masts gyrating wildly against the leaden sky. The M10 butted the sea under *Vindo*'s bow and we swung in towards the black, streaming sides. I remember thinking of a big, black whale breaking surface, as the counter stern swung across the seas to give us a degree of shelter.

'Looks a bit better now, son.' Jim cast an experienced eye at the lessening run of swell along her side and we turned in towards the short pilot ladder swinging over in readiness for the scramble aboard.

I remembered how I had listened respectfully to tales of pilots being put on board loaded steamers, actually stepping from the boat onto the side light screen on the bridge wing. I had always thought of such stories as being told tongue in cheek to impress young boys, but looking at this ship as she had been rolling made me begin to wonder. The combination of a thirty-five degree roll with a lateral slide down a swell, at the same instant a small boat being picked up by a confused sea, almost being dashed against the ship's bridge, a moment's opportunism, an agile leap, suddenly it was possible.

However, Old Jimmy was long passed the agile leap phase of life, if indeed, he had ever reached it, and here we were, praying for just fifteen seconds of comparative calm to enable him to be up that ladder, across that ten feet or so of side decking and onto the safety of the steel ladder leading to the main deck.

We watched and waited.

'Right! Now, son!'

I swung the canvas hood flat and crouched at the engine controls. Jim Powell stepped over the hood and stood for'ard, holding onto the short rope painter to steady himself. We roared alongside.

At the same instant the old steamer seemed to yaw round, running completely off her helm and an enormous breaking swell ran along her side almost at main deck level. Peter, seeing what was happening, just managed to get the tiller over as the wave hit the boat's transom and we felt ourselves being lifted bodily, with the ease of a ping-pong ball. There followed a jarring crash; the engine screamed its head off as I grabbed the throttle. The Pilot picked himself up from the coils of anchor rope, obviously very shaken, and looked around.

'We're ashore, son!'

Not a statement to be made lightly.

We were, in fact, high and dry on *Vindo*'s deck and in very real danger of sliding off sideways into an angry sea several feet below. We looked at each other in a state of shocked bewilderment. What to do? Leap out onto the steel deck and try to make that iron ladder or stay put and see what happened.

Out of the corner of my eye I glimpsed a burly figure, ankle deep in water, pluck the Pilot out of the boat and almost throw him bodily up onto the main deck. It was the Mate of the *Vindo*, a gigantic Estonian, with hands like a side of beef.

The next instant, we were on the move again as another vicious roll and a flurry of white water sent us slithering and screeching back into the sea again.

'I'm pleased about that,' said Peter, 'That big ship roll was making me seasick.'

'Yeah, and that Mate didn't even say, "Welcome aboard!"' I tried to stay nonchalant but was feeling very sticky in the underpants.

We checked around but didn't seem to be making any water and so set off back towards the cutter, all the time keeping a wary eye on the state of the sea.

Then the engine stopped.

I whipped open the engine box cover and tried to look intelligent. You could have fried an egg on the cylinder head! The cooling water outlet was hissing steam and everything looked very hot and solid.

'Engine's probably seized, we'll have to row her.'

Out came the oars and rowlocks and we began the onerous task of trying to row a very heavy, awkward boat, half full of water.

Meanwhile, back on board the cutter, the skipper had seen our predicament and was organising a tow rope and someone to take charge of the after deck. Jack, the next senior Pilot, sometimes known affectionately as 'The Bosun', and the junior Pilot, Matty, went aft, aided by the fireman who had emerged from the stokehold, looking like a half-clipped poodle but completely oblivious of the odd looks he was getting. The only piece of line long enough to use as a tow rope for us was very thin heaving line, the sort of thing you would see attached to a life buoy, with a relatively small breaking strain. This was going to be tricky. As the cutter approached us I unshipped my oar and stood by to catch the tow line. Peter concentrated on not getting the boat athwart the sea. At the second attempt I caught the line.

'Take two turns round the post, then one under the hook, and hold on to the end,' screamed the Bosun. 'Don't make it fast, if it comes up with a jerk it might part, so stand by to surge it round the post!' All sound advice.

Peter unshipped his oar and the line came tight with a TWANG, and the boat surged ahead as the cutter set off back inside. Each time the boat over-ran the tow rope as she surf-boarded down a swell, the two-man team on the cutter hauled in the slack as fast as they could and then, with two turns round the bollard, surged the rope with a high pitched ZIZZZ as the weight came back on it.

The steam cutter rolled and lurched through the heavy, confused sea at the lighthouse. Matty shouted, 'Don't look behind you, Peter,' but Peter, being a trusting soul, turned round to see a wall of breaking water eight feet high, bearing down on him. He went very pale.

Up for'ard I was having a marvellous time. By gum, this beats rowing, thought I, until I found I was holding the bare end of the heaving line between my thumb and forefinger! I had nowt left!

It was up to those on the cutter now, labouring away, serving that tow line. They stood there panting for breath, soaked to the skin, hands blue with cold off the wet rope, working like demons to ensure the thin line –

our life line – didn't part.

Eventually we were safe. The '*B.O.*' had run inside in sheltered water and, as they anchored, we were dragged alongside, shivering with cold and fright.

Fireman Jack, looking a bit worried, helped me out of the boat with a grimy hand, with, 'Will you have time to finish me off before you go home?'

'I don't know, Jack,' says I. 'But don't worry, I'll leave the gear and you can get one of the night watch to do it for you.'

He disappeared down aft, muttering to himself.

That night in the relief bus going home, I said, 'What's on tonight, then, Pete?'

'Don't know yet; what time does Evensong start?'

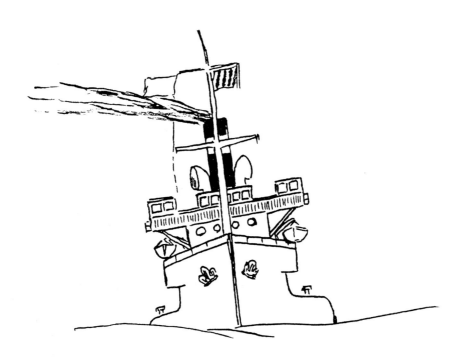

CHAPTER 9

Winter time for us was the period between October and April, although bad weather could come in the North Sea at any time. For instance, we often had a northerly gale in August just to keep us on our toes. January, February and March were usually the coldest times at sea as well as ashore, although, curiously enough, when there was a bone-cracking frost on the land and everything was crisp and even, very often by comparison it would feel much warmer on the water, especially if we were experiencing nice, calm conditions. The sea could be like glass with maybe a very slight, lazy swell, whilst ashore the foreshore and hills behind were white over.

In these conditions you could spot ships many miles away, often just a mast being visible, whilst the rest of the vessel was below the horizon or 'hull down' as we say. Sometimes it was possible to predict what ship it was we were sighting as we had quite a lot of regular traders who would arrive at roughly the same time every week. One such company who traded regularly into the Tees was 'Gibsons' of Leith, and they had a fleet of coasters which we knew very well. The little *Quentin* was one of these and would arrive very regularly from Leith and load in Middlesbrough for Antwerp.

One bitterly cold February day, the young Pilot on turn was John Mifflo, a powerfully built, extremely quiet man, who had not long since returned from sea service in the Merchant Navy. He was impatient to get aboard a ship and, in consequence, was keeping a vicious look out to the nor'ard. Eventually his patience was rewarded with a glimpse of a top mast right on track from the Firth of Forth.

We boathands, having other things to do, did not share in his feeling of anticipation. To us it was simply another ship to run off to in the motor boat. The day's routine was well underway with the two junior lads doing the menial and routine tasks, while Arnold, senior hand by six months, was flitting about the winches and steering engine with an oil can. Peter's turn to scrub out that morning meant I was polishing the brass work in the wheelhouse. Both were jobs we heartily loathed but they had to be done three times a week so we made the best of it.

The Pilots, sitting in the saloon, grumbled and muttered as Peter appeared at the door with his massive, galvanised bucket, ready to do battle, armed with a slab of carbolic and a scrubber. His sleeves were rolled up to the elbow and two large, wet patches in the knees of his old moleskin trousers indicated that he had been grovelling under the bunks in the sleeping berth below.

Peter's moleskins were remarkable trousers in that towards the end of their long career, they were the only pants I can truly say would stand up of their own accord without the internal assistance of their owner. They became so salt-encrusted that when they were dry you could hear them cracking.

Peter, once having evacuated the saloon, began to scrub with the vigour of a man with a purpose.

Up in the wheelhouse I was keeping an eye on the black smudge on the horizon getting nearer. It was one of those very still winter days when even the exhaust from a motor-ship's funnel becomes visible as it hangs about in the crisp air. A sparkling, white bow wave showed up well against the shining black of the hull.

At about four miles or so, big John could contain himself no longer, and shouted for us to man the motor boat which was lying alongside in the soft swell, gently nudging back and forth against the restraining bow and stern ropes. Arnold grabbed his jacket and, laying the oil can in the scupper, leapt into the boat. Peter stuck his head out of the cross alleyway door and shouted up to me in the wheelhouse.

'Can you go while I finish off this saloon?'

I dropped my mutton cloth and slid down the brass rail of the bridge companion way, grabbed my old gas cape and made for the boat. In the boat Arnold was cursing the engine; he had already skinned his knuckles painfully on the gear box cover, but the pawl and ratchet system of cranking the engine just wouldn't engage. We often had this bother; it was usually cured by a sharp thump delivered with the heavy end of the tiller on the offending part, but today it would have none of it. Either the pawl and ratchet were too badly worn to engage or else it was just too gunged up with countless oilings. Cranking the engine by hand was the only means we had of starting the old Morris Vedette. The powers that be had decided that electric starters were a needless luxury on boat engines, if indeed they were even available.

Try as we might – and we all had a go – we could not get the engine cranked into life. There was no alternative but to use the pulling boat. So we clattered around the steel boat deck changing the clutches over on the boat winch, making sure the pulling boat had all its gear in – including the plug – lifting it out of its chocks and then winding it over the side in its davits.

Someone yelled to Peter, up to his elbows in mucky water.

'Get him on the lamp, Pete, tell him to ease down and have a boat rope ready. He's comin' at us like a mail train.'

Peter emerged from the cross alleyway again, his hands, and especially his finger ends, all lily white and crinkled with being in soapy water too long. He sized up the situation at a glance and pounded up the brass-bound ladder to the wheelhouse. The sliding door crashed back against its stops as he grabbed the Aldis and began signalling for all he was worth, soap suds running down his arms and dripping off his elbows. Down below him we sheered off in the light pulling boat and began rowing an interception course that would take us to a point a few hundred feet ahead of the ship. Had he seen our signal? Could he reduce speed in time? Would he have a boat rope ready? All these things raced through our heads as we drew closer. Arnold had the starboard oar and his cousin John was applying considerable muscle to the other one. I was keeping out of the way as best I could as I didn't have the weight or muscle that the situation demanded. We approached as close as we dared under *Quentin*'s port bow. He had twigged the situation and, although very late in the day, was doing his best to make a seamanlike job of what looked like being a fiasco.

Yells from the deck.

A very Western Isles accent sent a hand racing for'ard to the bosun's locker under the fo'castle for a suitable length of boat rope. The ladder was rigged over the bulwark rail about fifty feet for'ard of the wheelhouse, but as we clattered into his port bow with the boat's shoulder, he was still doing a good five knots, and that was four and a half knots too fast. Arnold unshipped his inside oar and yelled for the boat rope to be thrown. Big John thrust his oar into my hand saying, in his gruff voice, 'Keep her sheered hard in while I get on board.'

'Crikey, surely he's not going to try for the ladder at this speed.'

John dashed over the thwarts and made a grab at the ladder as it arrived within his grasp.

A rope was thrown in desperation from the deck and wrapped itself round Arnold's head and shoulders. I got a glimpse of John with his arm hooked round a rung of the crazily swinging pilot ladder. I looked to my oar and pushed for all I was worth to keep the boat hard alongside.

Another glance for'ard – no sign of John – by gum, that was quick!

A hat went floating by.

A very wet arm came over the bow, followed by a big, red face, mouth gawping open like a cod fish, hair plastered down his face.

'Bear off,' yelled Arnold. ''E's in the oggin.'

We managed to clear the stern of the little ship and get away from the danger of a moving propeller. The Pilot was by now grasping the gunnel with both hands, utterly speechless with the sudden shock of the icy water. His uniform mac floated up around his waist as we tugged and heaved to get his considerable bulk over the side and into the boat. Eventually, with a struggle, we had him sitting in the boat, sopping wet, leaking from every lace hole and gasping for breath.

'Where – where – where's me hat?'

Back we rowed and managed to fish it out of the sea. He sat clutching it as we rowed him back to where the cutter lay at anchor, gently dipping in the slight swell. Of course, by now we had quite an audience, everyone was out on deck. John's uncle, who was also Arnold's father and one of the senior Pilots, yelled at him.

'Don't come back here. Get yourself aboard and dry out, then home as quick as you can.'

Poor John, shivering, nodded in agreement, and so once again we pulled over to the *Quentin* which, by now, was lying all stopped quite close by.

John was helped aboard by two grinning Shelties and as Arnold and I pulled away from the ladder I could hear slop, slop, slop, squelsch, squelsch, squelsch, as he shuffled along the deck, leaving great pools of water everywhere.

We heard, much later, of the skipper wrapping the Pilot in a couple of blankets and pouring red hot toddies down his neck all the way up the river. By the time they reached their berth, the shore gangs were amazed to see what looked like Chief Sitting Bull on the bridge wing, with a face a deep red, tinging on vermilion, swathed in blankets and roaring like a lion after the best part of a bottle of rum had taken effect.

Much later, a sorry figure in a very crumpled uniform dragged himself aboard the Seaton bus at the Transporter. Smelling quite strongly of seaweed, the blue of his coat mottled heavily white with dried salt, his shoes cracked and ruined and his socks sticking out of his pocket, a very inebriated junior Pilot surveyed the world through jaundiced eyes from under the broken peak of his cap.

'Fares please.' The chirpy clippy minced down the aisle of the double-decker and stopped dead at John. 'Ee! What's the matter, love? 'Ave you fell in?'

CHAPTER 10

One does not have to work in and around the world of ships and shipping too long before coming into contact either directly or indirectly with a distress situation. More often than not it does not involve you personally, but very occasionally it does.

For several days we had been hearing vague reports about a ship called the *Irene L*. She was on her way but was having some engine trouble, and her ETA was going back and back. This was not an uncommon thing as ancient Greek steamers did have a reputation for getting into one kind of trouble or another. Propellers were known to drop off. Sometimes, if they ran into prolonged bad weather, they would run short of fuel. They would run out of drinking water and food, and sometimes they wouldn't have enough ropes on board to moor them if someone had sold them at the last port! They were as often as not run on a shoestring, and shoestrings are always breaking. So for a ship of this type to lose a couple of days here and there was not too serious.

However, information filtered through from various channels that she had effected temporary repairs and was once more making for the Tees under reduced power. The weather was good, with no wind, overcast skies and a lazy but persistent northerly swell.

We had been anchored at the Tees Fairway Buoy all night and although reasonably busy had managed all our work from the motor boat. The skipper and chief engineer had been turned in since about one-thirty and we had lain reasonably quietly, except for continuous rolling, as the old cutter lay athwart the monotonous swell.

About four in the morning, just after I arrived on deck for the four to six watch, I spotted two mast head lights right end on to the north. A ship eight or nine miles away was heading directly towards us. I called him up on the Aldis lamp but got no reply, so I went back to the galley to start my morning's chores.

Biff, the younger of the firemen, wandered along to make a pot of tea and we passed the time by yarning about his various romantic escapades, real and imaginary.

Biff, we regarded in some awe as a man of the world, a great lover, a poor man's Casanova. He used to tease us about our constant state of being flat broke, while he, although always down to his last fiver, invariably had something going for him.

The grey, pre-dawn light gave way to a clear, overcast morning. The ship to the north was becoming visible as a steamer end on and coming directly towards us, albeit very slowly. I poked my head round the saloon door and informed the Pilot on turn, who was having forty winks on the settee.

'What is it, son?' Big Bill Braithwaite, a mountain of a man with a terrific Roman nose protruding from a big, fleshy, red face, arose from the settee; his blood-shot eyes blinked as they peered at me out of the gloom.

'I think it's that old Greek,' said I. 'It's come up very slow and won't answer the lamp.'

'What time is it, son?'

'About six o'clock.'

'Oh, great,' muttered Bill, 'and me all set for York races today. I won't get home 'til dinner time if it's that bluddy thing!'

He heaved his great bulk off the settee and staggered over, against the roll of the cutter, to the corner where there was a minute wash basin. Splashing some water into his face he turned to me.

'Call your mates, son, and get the boat ready.'

I nipped out of the cross alleyway and made my way round the deck housing to the entrance companion way of our fo'castle accommodation.

Arnold's voice came up from the darkness below.

'Aye, OK. I'm awake, I'll give Pete a dig and you two can go off while I cook my breakfast.'

Arnold had the job weighed up.

Sure enough, about five minutes later, they both appeared, Arnold in sand-shoes without laces, with an enormous plate of fry-up in his hand, and Peter, yawning his head off, pulling on his oilskins. Peter staggered as the cutter rolled and heaved, straining at her anchor cable, and kicked a galvanised bucket half the length of the deck as he lurched to regain his balance.

'Everyone's awake now,' said Arnold, as the bucket clattered along the cemented scuppers. He grabbed the big, black iron frying pan and tipped his breakfast unceremoniously into it. Whatever it was, it almost filled the pan.

Biff, who had disappeared aft into the engine room, reappeared carrying an enamel plate with bacon, egg, black pudding and sundry other items. He clip-clopped along the deck in his heavy, laceless boots, and stepped into the galley as Arnold was riddling the stove with a long poker.

'Blimey, are you going to eat all that at one go?' said Biff, eyeing the great heap in the pan.

'Certainly,' said Arnold in a huffy voice, 'and if you're thinking of frying that, you can forget it for at least half an hour. This stuff has to be heated slowly.'

Biff decided to settle for a half a pint of tea while he waited.

Meanwhile, Pete and I and Big Bill half jumped and half fell into the waiting boat and while Pete swung the starting handle Bill let go the bow line and I slammed the tiller into its brass socket. The engine roared into life and off we went. Once in the boat and on our way, it became apparent that this ship was steering a remarkably good course; in fact it had hardly wandered more than a degree or so since I first sighted it. It was also becoming increasingly apparent that it was heading directly for our cutter, lying unsuspectingly at anchor a mile ahead of it.

As we approached closer in the improving light, we could quite clearly make out what appeared to be a chaotic jumble of activity right aft on the poop deck. Fathoms and fathoms of wire lay strewn about the deck. Bleary-eyed, unshaven men were manning a massive docking winch on the poop which was clattering away, belching steam from every gland. Blasts from a pea whistle signalled from the bridge sent men scurrying around the winch, heaving away on one drum while others, on the opposite side, slacked away on theirs. Over the stern of the ship, half hanging and half floating in the water, was a long spar about thirty or forty feet long, and at its very end were lashed two or three heavy hatch boards.

The ship had clearly lost its rudder and was making port with a makeshift jury rig.

Big Bill looked at us and we at him.

'Are you thinking what I'm thinking?' he said.

We were.

Directly ahead of the *Irene L* lay the cutter, now about half a mile away.

'Put me on board quick,' said Bill, 'And then get back and get that cutter underway.'

As we swung alongside the ladder, he looked up at the swarthy, tired face that peered over the bridge wing.

'FULL ASTERN!' He yelled so hard that the mouth belonging to the dark face dropped open with surprise.

'DOUBLE RING FULL ASTERN, AND HARD ASTARBOARD,' he yelled again.

For a big man he fairly skipped up the short ladder and dashed up to the bridge. We clearly heard the rattle of the telegraph wires from bridge to engine room, as we opened the throttle with a roar and tore off to warn the cutter.

At that moment there was a thunderous blast of steam and water from the ship's whistle as Big Bill heaved with all his might on the whistle lanyard. A plume of white steam engulfed the top of the funnel and then suddenly stopped as the lanyard parted under the strain, dumping the Pilot heavily on the seat of his pants, the raggy rope end still in his hand.

'Crikey Mick! He's going to run the cutter down,' and we began shouting and waving our arms.

Unfortunately, the way the cutter was laying to the swell, the galley door was slightly obscured to us and so we jumped up and down to no avail for quite some time, then a head appeared around the corner of the door. It was Arnold, wearing a puzzled expression.

We pointed at the ship, which was out of his line of sight, obscured by the deck house, and yelled in unison, 'LOOK OUT!'

Arnold ambled down the deck holding a piece of fried potato on the end of a fork, gently blowing on it to cool it. Turning the corner, an enormous bow and a pair of anchors stared down on him. He turned and ran for all he was worth, yelling to Biff at the top of his voice as he ran forward to the windlass, 'SET THE ENGINES FULL AHEAD.'

Peter turned to me in the boat as I sheered off and throttled back.

'Aren't you going back alongside?'

'Not likely, we'd better stand by here in case we have to pick up survivors.'

By this time the engines of the *Irene L* were going in reverse but to no great effect and she crept nearer and nearer the stern of the cutter. Arnold was dashing back and forth along the full length of the deck, one minute taking a quick look to see how close the vessel was and then back for'ard to give another turn on the already wide open steam valve, as the anchor cable came rattling in over the gypsy. Down below, Biff had managed to get the engines going ahead and by now everyone else on board the cutter was appearing on deck wondering what the racket was, some of them clutching life jackets.

Finally, the old steamer came to a halt as the propeller churned the sea to white foam under its stern, the great bar stem just missing the cutter's after decking by a couple of feet. In fact if they had dropped the anchors, they would have landed on the cutter's deck. Everyone heaved a great big sigh of relief at a very close call. We had all heard of pin point navigation, but this was ridiculous!

As the stricken ship sluggishly backed away from us, we could just make out two separate smudges of black smoke rising in the air from behind the line of the breakwater, and then into sight came a pair of tug boats, steaming full speed in our direction, black smoke belching aft from their funnels. As great creaming, white bow waves climbed high up their bows, they leaped towards us like a pair of great, black dogs, each carrying a bone in its mouth.

Bad news travels fast and bad news in the shipping world is often good news for the towing business. The magic word 'Salvage' brings instant response from tug skippers and their crews. A share in the prize money is a big incentive for men who work long hours for relatively small gain. So here were the tugs without whom it would be impossible for this hampered vessel to manoeuvre in the confines of the river and eventually berth somewhere where effective repairs could be made.

After much shouting, gesticulating, cursing and general pandemonium, they made fast, one for'ard and one aft, and slowly, ever so slowly, the old ship started her hesitant way towards the lighthouse. Now and again she would take a great lunge either to one side or the other, but each time a blast on the Pilot's pea whistle would send the tugs laying way out on their towing wires to counteract the sheer.

The Pilot dared not use more than 'dead slow ahead' on the main engines in case the old ship took charge of the situation and the tugs could not hold her. The last thing anyone wanted was for her to run aground and maybe block the main channel to other shipping. So, in the words of the Chinese proverb, 'Softlee, softlee, catchee monkey', or 'More haste, less speed', it amounts to the same thing in the end – and Big Bill didn't get to York races.

Back aboard the cutter, Peter and I were being ribbed mercilessly for our 'presence of mind' in lying handy in case it had come to the crunch.

Biff was back in the galley, quite unruffled. He lifted the black pan off the galley stove and, observing the equally black contents with feigned distaste, asked Arnold if he still intended eating it. The burnt mess went over the side and the pan given a rudimentary wipe before being returned to the stove with the fireman's three rashers of bacon lying almost smugly in the bottom of it.

Arnold came aft after screwing up the brake on the anchor.

'How about a slice of dip, Biff?' he asked, almost sheepishly.

'I might think about it,' said Biff, 'But you'll have to wait, I don't like my bacon doing too quickly.'

CHAPTER 11

Spending our working apprenticeship days in and out of boats and generally being part of the scene that existed in those days in the river estuary, we came into contact with many colourful characters. Some were chaps who would finish work at ten o'clock at night at the local steel works and then come straight down to the breakwater and fish until two or three in the morning. We used to think they were crazy. A lot of men working in the nearby steelworks owned boats. These were invariably kept in a rough haven or tiny harbour fashioned out of great lumps of broken slag. It was this slag that formed the bulk of the South Gare breakwater which had been completed, largely by using Irish labour, in the 1880s. Because of this Irish labour, and the fact that they lived in little shanties erected around the tiny boat haven, it came to be known as Paddy's Hole, a name that has stuck to this day. The shanties have long since gone but in their place has sprung up an assortment of fishermen's week-end huts. Some are quite elaborate affairs, albeit self constructed, and comprise miniature slipways, sun verandahs, even a small dry dock.

A growing number of men spend their entire weekends messing about in boats away from the wife and kids and, curiously enough, while the Paddy's Hole people have prospered and multiplied, it has never become commercialised like so many other little 'secret' places. Maybe it's the backdrop of steel works and oil storage tanks that has not attracted the artistic fraternity or the 'fashionable' people.

One of these breakwater characters was Rudolf. He was, as far as we could make out, an old Estonian or Russian, who lived with his considerable family in a derelict house-boat, moored up a quiet backwater we called the Snook Channel. He was a real hard-bitten character, fiercely independent, always on his own, his great calloused hands with their broken and split fingernails grasping the heavy wooden oars as his powerful, bent shoulders propelled the massive double-ender he rowed. The way he managed without the help of an engine was simple. He always rowed out on the ebb tide, anchored and fished over two tides then rowed back up on the flood. That way, the currents did most of the work for him, but the only snag was that if the weather suddenly deteriorated he was in a very vulnerable position, and several times we came to his rescue and towed him out of trouble. I think he appreciated us keeping a weather eye on him but of course would never say so, and we were too discreet and fond of him to ever suggest that we should 'rescue' him. He just accepted our tow rope with a cheery wave and a smile on his stubby-bearded face. As often as not, when we dropped him off with just an easy row back home, he would wave us alongside and then hurl a couple of big codling in our boat from the silvery pile that lay inert in his tarry bilges.

'Tanks a lot, lads,' he would yell in a thick Baltic accent. 'Old Rudolf pays his vay,' and then he would lay his back into those long, heavy, sweep oars, his woolly hat pulled right down over his ears as the rowlocks creaked with each stroke.

I think he had a great affinity for the cod and, looking back, I can most often picture him in calm, bitterly cold weather, the kind that seems to encourage good cod fishing. Many times we would see him pulling to his favourite spot quite near the lighthouse, and, in the gathering gloom of a winter's afternoon, he would be seen anchoring and arranging his hand lines at various points around the boat.

When darkness fell he would light a battered hurricane lamp and keep it standing in the bottom of the boat so that it would illuminate the interior with an eerie soft orange glow. His silhouetted figure could be seen moving about tending his lines, with the odd flash of silver as another prime codling came 'in out of the wet'.

This would go on all night long for twelve or fourteen hours if the fishing was good. Sometimes during our night sorties, we would pass close to him and give a wave, or, if he wasn't to be seen, we might creep alongside and peep in to see him fast asleep in the bottom of the boat, with an old sack as a pillow and covered with an raggy oilskin, already sparkling and shimmering in the hard, pre-dawn hoar-frost.

'Are you OK, Rudolf?' we would ask if he opened an eye.

'Ya. Ya. Go avay, you frighten ze fish!'

We would motor off quietly, full of apologies for disturbing his sleep. Rudolf was truly a man of the sea.

Maybe not quite so hard but, nevertheless, just as much a character, was Spot. We all knew him as Spot and indeed it was years later that I found out his real name. He seemed to be permanently employed in ferrying Pilots who lived on the other bank of the river at Seaton Carew to the Pilot Station at the South Gare.

The great majority of the Service lived in the Middlesbrough area and so, as the Pilotage Service was run from a base at the South Gare breakwater, everything was geared up to reliefs and watch changes being done on the south side of the river, the Yorkshire side. The handful of Redcar men also found this arrangement very convenient, as they lived only four or five miles down the coast in the seaside town. Not so the Seaton men. If special arrangements, like Spot's boat, could not be made, they faced the long and tiresome journey right round from their little town on the Durham side of the river, first to Port Clarence, over the Transporter Bridge, then by train or bus to Redcar, where they joined the relief taxi for the final trip down the breakwater road. This journey, with missed connections and other annoyances, could take two hours, especially frustrating as the breakwater, government jetty, cutter and shore cabin are clearly visible a couple of miles across the water from the Promenade at Seaton Carew.

Spot was a diminutive man of little more than five feet. I never saw him in anything but thigh boots, an old navy-blue coat which had been plastered with a concoction of linseed oil to waterproof it, and a black cloth cap. He wore his cap constantly; I suspect he never took it off willingly. I once saw it blow off and it revealed a round, shiny, lily-white pate, in grotesque contrast to his weather-beaten face, as wrinkled and brown as a walnut. He was so mad at losing his cap over the side that he grabbed it from the water, banged it twice on the boat's gunnel and rammed it back on his head. Freezing cold sea water ran down his face. 'Hee, hee, hee,' he chuckled at our obvious distaste, 'It'll soon warm to the skin!' And of course he was quite right. Clothing soaked in salt water does warm to the body. If that wasn't true, we, as lads, would have been frozen stiff most of the time. By the same token, clothing soaked in salt water never dries out properly; the water content may be evaporated but the salt remains and, of course, salt attracts water so as soon as conditions become the slightest bit damp, your clothing feels damp again. The only cure is a good washing in fresh water to get rid of all the salt, but I don't suppose Spot worried about a little detail like that.

He pottered about, day in, day out, in all weathers in an old pilot boat that had survived for nigh on a century from the days when each Pilot was in competition for work. Each would have a boat to launch from the beach, and a lad, usually an apprentice, and together they would go 'seeking' for miles. When they found a ship bound for the Tees, the Pilot would strike a bargain with the Master and either leave the lad to sail back as best he could or tow him back astern of the ship. This system was known locally as 'doggin' and it survived up to the beginning of the First World War.

Spot's boat, the *Mary*, had long since been converted to power by the fitting of a lusty, single cylinder engine, and the stern was ballasted down with sacks of pebbles to keep the screw in the water, but you only had to cast a half-knowledgeable eye on her to see the beauty of line and form of a sailing boat. Indeed, it was often said that in her day she was the fastest doggin boat under sail on the coast. Spot was very proud of her and the way she could speed along; her one lung banging away merrily was audible for miles in quiet conditions. Spot was such a regular face around the place that once, when he went missing for about a week, Pilots and cutter crew began wondering what was wrong. The explanation, when it eventually emerged, caused some amusement.

A Canadian Liberty ship had arrived in the bay and anchored and, while this in itself was not unusual, the Pilot he had requested to supervise the anchoring, came back on board the cutter with tales that made everyone's eyes sparkle.

The ship was a veritable Aladdin's cave of all kinds of goodies that we had not seen since before the war. The Chief Steward had taken him down in the massive store room and shown him sides of bacon, tins of ham, salmon, fruit, sacks of sugar, white flour, sweets, chocolates, silk stockings, cotton and silk poplin shirts and pants, bed linen, blankets and towels and innumerable other items that were either rationed or scarce or both. Apparently he and the 'old man' were selling the lot as the ship was changing hands and everything had to go.

Well, as you can imagine, the excitement was intense. Money was produced from all sides, some from the most unlikely sources, and two of the young, senior Pilots, Ephie Dee and Frank Crimdon, set off with quite a roll to do business. They seemed to be away for hours. Those of us left on the cutter kept a watch with the glasses on the boat moored alongside her ladder. Every now and then someone would appear on deck with a well stuffed sack and it would go over the side into the boat. This went on for some time until, finally, the two Pilots appeared, both glassy-eyed and a trifle unsteady, each smoking an enormous cigar. By this time it was getting very near five in the evening and relief time, so the skipper was anxious to up anchor and get inside to relieve the watch. Back came the boat, loaded down to the gunnel, with all kinds of parcels and sacks which were quickly whisked into the saloon, the boat picked up, and the cutter steamed inside. In the saloon, five or six Pilots were having a heck of a time dividing up the spoils into more manageable parcels. There were tins of butter wrapped in towels, half sides of bacon in pillowcases, sacks of flour and sugar dumped in various corners and tins of sweets and candies strewn everywhere, with two-pound slabs of chocolate cluttering the table. To make matters worse, the two unfortunates who had bought it all couldn't remember what they'd bought and, what was worse, how much they'd paid. Bottles of honey got mixed up with bars of soap and there was even a carton of sanitary towels, and a box of laxatives that had obviously been stuffed in surreptitiously by the wily steward.

There was hell on.

As we dropped anchor close to the jetty, it was beginning to sound like Stockton High Street on a market day. Into the middle of all this mêlée sailed the ever faithful Spot with a boatload of relieving men. Now Spot, as far as I could make out, didn't smoke or drink very much, but he did have a very sweet tooth, and it appeared that someone amongst the general hubbub, in a sudden fit of generosity, gave Spot a big slab of dark chocolate. This delighted Spot who thrust it into one of his less raggy pockets and, chuckling with glee, shouted 'Thanks, Mister,' as he opened the throttle, pushed the tiller hard over, and was gone.

Next morning, no Spot.

In fact he was missing for several days without explanation, until in the course of a casual conversation, Barney, one of the more mischievous Pilots, happened to mention that the last time he saw Spot was when he gave him the two pound slab of chocolate.

'Blimey,' said Ephie, 'Was it light coloured like milk chocolate, or very dark?'

'I think it was dark,' said Barney, 'Why?'

'You bloody fool, that was laxative chocolate,' Ephie said, and then started to laugh. 'I only hope he read the wrapper before he ate any of it.'

'Well, if he didn't, at least we know where he's been all this week,' Barney giggled.

When Spot appeared some days later, looking a bit pale and thinner, we asked him how he was.

'Eee, Mister, I've had a terrible time with the runs, didn't dare cough for four days. I must 'ave an arse 'ole like a Flanders' poppy. It's been a bug I've picked up somewhere.'

Nobody dared tell him about the chocolate.

CHAPTER 12

Over the period of four years we spent as apprentices we had quite a few changes amongst the crewing of the cutter. With two minor exceptions we had the same two skippers for the whole of our time. The elder was known to everyone as Captain Blood, and the younger, though senior in service, was Captain Jack. The engine room staff was a different kettle of fish. Firemen came and went, some almost daily, and we had quite a few engineers although they seemed to stay a longer period. The Cutter Company seemed to have great difficulty in replacing engineers when they left for various reasons and replacements were sometimes a little odd.

One morning, a stocky little chap appeared complete with brown paper parcel under his arm and carrying an ex-army haversack. He had a shock of black curly hair and thick, pebble-lensed glasses which made his eyes look very small, but by far his most striking features were his hands. For a small man he had the biggest and most ungainly-looking hands I had ever seen and he seemed to carry them in front of himself as if on display like two great bunches of bananas. They were very white and the backs of them were covered in coarse, black hair. From all accounts he had managed to be sea-sick in Spot's boat on the way down from Graythorpe. Considering it was a calm, sunny morning we weren't giving much for his chances.

He spent the best part of three-quarters of an hour with old Ernie, the engineer going off duty, before the same Ernie departed declaring himself happy about the new man's competence.

The morning passed uneventfully and apart from the odd bang or loud hiss from the engine room, everything seemed fairly normal. Ships' engine rooms are seldom quiet places and ours was no exception but I must say our new engineer or 'The Gook' as we quickly nicknamed him, seemed to have found quite a few new rattles, bangs and hisses that the rest of us had never heard before. At lunch time he arrived at the galley door with his spotless white boiler suit newly scorched on the seat of the pants and expressed his intention of mashing a pot of tea. His hair was neatly parted still but his face covered in soot. A sweat rag encircled his neck with the end in the corner of his mouth.

Having made his tea he announced in a loud voice that he wasn't entirely happy with the main generator and that very afternoon he intended to do some minor but highly technical adjustments to improve its efficiency. At this the skipper, Old Blood, looked somewhat alarmed and disappeared down below to his cabin shaking his head and muttering to himself.

About 2 p.m. all the lights went out.

Down below 'The Gook' was hopping around in the darkened engine room, his right hand clamped firmly under his left armpit, cursing and swearing in a loud voice. The fireman was first on the scene, almost landing on top of the engineer as he slid down the iron ladder in the darkness.

'What's up, Chief,' the fireman asked respectfully.

'I turned the steam off the genny, but it ran so long that I thought I'd slow it down by grabbing the fly wheel – I think I've broken my finger!'

Everyone crowded round the poor Chief who was obviously in great pain.

'I think you'd better get it x-rayed and examined properly at the hospital,' said Captain Blood who had had his afternoon siesta disturbed by the commotion. 'We'll send for a taxi for you.'

So off went the luckless engineer nursing his great, pale hand Napoleon style in his jacket.

A couple of hours later he arrived back and stood on the end of the jetty with his arm in a sling looking very self-conscious. We went to collect him in the motor boat and he just managed to climb down the vertical iron ladder into the boat.

'We thought you'd be going straight home, Chief,' Peter said.

'Hell, no,' said the 'The Gook'. 'If I had to get Ernie back today he'd skin me alive. I'll have to stick it out 'til the end of the watch. Besides, it's only a broken finger and it's all in plaster so it won't come to any harm.'

In plaster it certainly was. All his arm from the elbow to the tips of all his fingers. Only his thumb remained mobile.

'How are you going to work the engine with only one arm?' Pete persisted.

'Dead easy, son. I always said I could do the job with one arm tied behind my back.'

Peter looked at me and I at him. This was going to be worth seeing.

When we got back to the cutter Old Blood announced that we'd get under way and steam out to the Fairway Buoy and anchor on our sea station after the oncoming night watch had had their tea.

'Are you sure you can manage the engines, Chief?'

'I'll be OK, just take it easy 'til I get the hang of it.'

The watch of pilots duly relieved at 5 p.m. but we, the lads, were working right through and doing a 24-hour shift until eight the following morning. After our tea Peter and I couldn't resist a look down the engine room as we could hear the familiar sound of the triple-expansion engine being warmed through.

The engine controls on the manoeuvring platform consisted of an overhead steam valve or throttle to control the speed of the engine and a large vertical polished steel wheel about two feet in diameter which had to be spun very fast in one direction or the other to control the direction that the engines ran, either ahead or astern. This could also be done if one started the wheel spinning first by a short, vertical, steel lever which directly controlled the position of the cross-head coupling. A degree of skill and experience was necessary to throw the cross-heads and thereby manoeuvre the engine smartly ahead or astern. All these controls, and indeed the whole engine, were burnished and shone like silver. We sometimes watched the engineers at work when we could and became quite critical of the skills involved as they worked below coaxing the steam valve, standing clear of the wildly spinning wheel as the crosshead couplings clattered across reversing the engine direction, and of course answering the telegraph from the bridge while all the time keeping an eye on the steam pressure valve and gauge glass.

The practice of warming the engine before any movement was undertaken consisted of gently rocking the engine ahead and astern to get rid of any accumulation of water or oily sludge which might have built up during a period of lying idle. The steam valve was just cracked open and the engine allowed to turn gently first ahead and then astern and then ahead again, just half a turn or so each way on the propeller shaft, but of course this meant the cross-heads being constantly clattered across one way and then the other and the wheel control being spun quickly one way then stopped then the other. This went on for as long as the engineer deemed it necessary before signalling back to the bridge via the telegraph that he was ready to go. The skipper rang 'stand by' on his telegraph and the engineer answered by ringing 'stand by' on the engine room telegraph when he was happy.

This then was the scene we were observing as 'The Gook' tried to cope with his mildly rebellious charge, with one good arm and one gammy one. The fireman, we noticed, was keeping well out of the way, busying himself in his stokehold feeding the twin furnace doors with great shovels full of best Welsh steaming coal.

The door to the cross alleyway slammed shut and Old Blood, still muttering to himself, climbed the bridge ladder clutching a pint pot of steaming hot tea, opened the wheel-house door, stood his pot down on the battery box and rang the telegraph.

The sudden clanging in the engine room two feet from his head just about made the little man jump out of his skin. The fireman, I think more out of curiosity than anything else, came through from the stokehold and answered the telegraph at a nodded indication from the Chief.

'Heave away,' Old Blood shouted from the wheel-house window, and Pete and I scampered along the deck and began heaving up the anchor.

We could hear the skipper singing loudly to himself some ancient ditty in a toneless basso profundo. He often insisted these tuneless dirges were real sea shanties but nobody I knew recognised any of them, not surprising really. Anyway it indicated that his humour had improved so it suited us.

Once up, the anchor windlass was screwed up tight on the brake and we shot down aft to spy on the engine room staff again down below through the open gratings.

As it happened, the way the cutter had been lying at anchor, between two inner channel buoys, necessitated a considerable amount of 'backing and filling' or alternately full ahead, and full astern movements, and I suspect a spark of mischievousness lurking darkly under that old black beret the skipper always wore. Anyway, he stood on his bridge quietly smiling to himself and blowing contentedly on his hot tea while rattling the telegraph vigorously back and forth ahead and astern. The effect in the engine room was both comic and chaotic. The fireman answered each call with a flourish of unnecessary telegraph twirling while almost doubled up with mirth at 'The Gook's' antics.

He was leaping about like a Dervish. Grabbing the steam valve, heaving on the cross coupling wheel, shutting the throttle, leaping back into action as the telegraph jangled. Whipping open the steam throttle, spinning the great wheel, getting it stuck half way, juggling with the cross-heads lever, only to get it going the wrong way. Stopping. Starting again. Mopping his brow with his lame arm as the sweat began rolling off him. FULL ASTERN. Close the valve, wait 'til the engine almost stopped, spin the wheel, clatter clatter clatter, open the throttle. FULL AHEAD. Shut the throttle, and so on. He got to the stage where he was opening the throttle instead of closing it, getting it going ahead instead of astern and vice versa, and all the time dancing around on the greasy footplate. How he managed to stay on his feet and avoid falling into the great flailing connecting rods that were pumping away barely a yard from his head, was a minor miracle. Eventually the skipper, having turned the cutter to his satisfaction, left the telegraph at full ahead and we steamed out majestically past the lighthouse.

The engineer leaned back on the ladder handrail, a great big grin of relief spread across his face, wiped his steamed-up glasses with his sweat rag and gave a thumbs-up sign with his thumb sticking from the plaster cast. The fireman grinned back and disappeared into the stokehold to attend his furnaces.

The cutter described a wide arc to starboard after passing the fairway buoy and after a couple of minutes of full astern, down went the anchor. Blood rang 'finish with engines' as the old craft came up to her cable and we screwed up the brake and popped along to see what the situation was like in the galley.

In the saloon the night watch was going through the orders that had come down at five in the tatty old leather bag from the Pilot Office. These consisted of information from the various shipping agents in the town concerning the expected times of arrival (ETA) of ships bound to the Tees. Sometimes it would include packages of crews' mail, sometimes just a simple postcard with the date, ship's name, agent's name, berth and ETA. It gave us a good indication of how busy we were going to be during the night. The cutter itself had no radio and so we relied heavily on this kind of information.

The fireman met us in the galley, still chuckling to himself.

'Poor old Chief,' he grinned a great toothless grin. 'He's having a rough time of it.'

'What's the matter now?'

'All I said was do you fancy a bit of bacon and egg for your tea? and he went a funny colour and disappeared into the toilet.'

'Do you think he'll be all right to keep a watch?'

'Well, I'm turning in when I've had my tea, you can ask him yourself when he comes on deck,' and with that remark the fireman departed, with his tea, down into his own quarters right aft.

As the lights gradually faded from the sky and the shore lights winked and blinked at us across the water, a serene quietness descended on the ship rolling gently in the oily swell. The little breeze dropped away completely and all the stars shone out with a precision and clarity that almost made you think that the great constellations had been arranged on purpose for our convenience and pleasure. I often amused myself at these times by identifying as many major stars as I could, sitting in the darkness of the wheel-house. The only sounds were the occasional clack of the anchor cable, the surging splash and lingering dribbling noises of sea water being displaced by the turn of the bilge as the old cutter rolled. Occasional laughter filtering up through the ventilator cowl from the saloon below. Bellatrix, Betelgeuse, Rigel, Sirius, the feed pump started banging away in the engine room and then stopped again, Capella, Aldebaran, Mirfak, Procyon, what names to conjure with. Bang! Bang! Bang! The feed pump started up and stopped again.

Then I noticed the little Chief moving about on the boat deck with a flash light. He appeared to be unscrewing the main tank filler cap. The Chief disappeared again and I resumed my constellation spotting. Bang! Bang! Bang! This time the pump didn't sound quite the same as usual. Looking aft across the boat deck everything seemed hazy abaft the funnel. Strange.

The engineer's flash light shone through the haze, gyrated about like a search light, and went out again. I went out on to the bridge wing, walked aft across the boat deck and only then realised that the whole of the after end of the cutter from the funnel was enshrouded in thick mist. How very odd! I looked round the horizon, it was clearly defined. Then I realised this was steam. Where the hell was it coming from? 'The Gook' appeared out of the mist, his torch blazing away in the eerie atmosphere and, diving into the engine room door, immediately vanished into the gloom. I climbed down the short, vertical, iron ladder onto the main deck and made my way to the engine room door. Peering in, I was almost knocked flat by the Chief coming out! The visibility in the engine room was about two feet and getting thicker by the minute.

'What's up, Chief?' trying to sound casual.

'I have a slight problem. Soon have it fixed.' The Chief peered at me through the water droplets condensed on his specs. His curly hair was plastered to his forehead in tight little ringlets and his boiler suit hung sodden and limp from his shoulders.

'Shall I tell the skipper about all this steam?'

'No, it's OK! I'm just going to have a look at the blue prints. I've got some sorting out to do with the main feed valve. It shouldn't take long. If anybody asks you where I am – I'll be down aft.'

'OK, Chief,' said I, and with that he disappeared again.

Just then one of the Pilots came out on deck for a breath of air and, hearing the constant hiss of escaping steam, turned to see the whole of the stern of the cutter enveloped.

'What's going on?' he asked.

'I think the Chief is having a problem.'

'What sort of a problem?'

'Well, he doesn't seem able to find some valve or other.'

'He doesn't seem to be short of steam anyway,' he laughed, and then, turning, he shouted into the cross alleyway, 'Hey, come and have a look at this, the Chief's having a Turkish bath!'

Within minutes everyone was out on deck and I was sent to find the Chief. I found him pouring over drawings and blue prints of the engine room on their table in the aft accommodation.

'You'd better come up, Chief, they all want to know what's going on.'

'Oh blimey, OK, son, I'll be up in a minute.'

With that the voice of Blood bellowed down the companion way.

'Are you there, Chief, what the hell's going on up here?'

The Chief shot up the stairs and out onto the after deck. By now the whole of the after deck was in a thick mist and everyone was milling around and bumping into each other.

'I can't get any water into the main boiler,' said the unfortunate Chief. 'The water in the gauge glass has gone right down and doesn't register any more and the steam pressure is rising all the time.'

'Do you mean we're about to blow up?' said Old Blood in an incredulous voice.

'Well, I hope it won't come to that. I'll have to let the fires out and cool everything down.'

'That means we'll be stuck out here without engines,' said Blood, 'like hell you will.'

The little Chief mumbled something about valves and feed pipes and dived back into the swirling steam with his flash light blazing away in the gloom. Meanwhile one or two of the Pilots appeared on deck wearing life jackets. One said, 'Let's all go ashore in the motor boat.' This brought a very severe look from Blood who was not amused at having his evening's rest ruined.

Finally Steve, the fireman, appeared, grinning as usual his toothless grin, and suggested that if the Chief couldn't get water into the boiler then why not go alongside the jetty and put the fresh water hose directly into the boiler from the shore line. Everyone thought this a good idea so we got underway and slowly limped inside.

Just to make sure that there were no mistakes made and perhaps to rub it in a bit, Old Blood had me call up the signal station on the lighthouse with the Aldis lamp and arrange for the other engineer to come down from home by taxi to sort the situation out.

It was a very irate Ernie who arrived at 2.30 a.m., having been dragged out of bed by the police. He stormed down aft with a face like a thunder cloud. Within minutes the steam had evaporated and calm was restored.

CHAPTER 13

Routine work for us boathands was very humdrum and we jumped at the chance to tackle a job that we hadn't done or seen done before. After all we were heartily sick of chipping and scraping the rust off bulkheads and bulwarks. Red leading and painting wasn't too bad because it was soon done and sometimes one of the Pilots would grab a brush and lend a hand. The Pilots themselves hated us to be clattering away with chipping hammers as the noise rang throughout the little vessel; there was no escaping it.

So when Captain Jack suggested that we flake out all the anchor cable on deck to take all the kinks and turns out of it, we all turned to with a good deal of enthusiasm.

Anchors and anchor cables were things we used constantly and they had to be inspected regularly for wear and tear. The anchor cable was stowed in a cable locker right in the forepart of the ship immediately forward of our fo'castle accommodation. The cable was made up in lengths of 15 fathoms called a shackle. Each shackle was joined together to the next one by means of a special link which could be broken. We carried five shackles on each anchor which was a total length of 75 fathoms of cable on each. The final end of the final shackle or the 'bitter end' as it was referred to, was made fast to a ring on the collision bulk head by a wire lashing.

As the first shackle was constantly in use it followed that it became more and more worn, so much so that the little studs which divided each link and supposedly prevented the cable from getting into a great unmanageable lump, often worked so loose that they would sometimes fly off the cable like bullets when the anchor was dropped in deep water in a careless fashion. To even out the wear, about once a year we would turn the complete cable end for end. This entailed hanging off the anchor not in use, removing the cable from the anchor by driving the pin out of the joining link and then, by means of long handled hooks, we would lift the cable off the gypsy and proceed to haul it down the deck. One of us would be sent down through the manhole into the cable locker to feed the cable up through the hole in the deck called the spurling pipe. This was a dirty, smelly job as the cable locker had quite an accumulation of evil smelling black mud and rotting seaweed. Eventually, after much heave ho'ing and a few bruised fingers, the whole of the cable would be lying out on deck in rusting rows like old cobble stones, wet and shining. The man in the chain locker then had to clean up as much mess as possible and pass it up in buckets to be dumped over the side. Usually at this point we would be sent off to make a cup of tea before the humping and heaving of stowing it all again began.

The first thing on re-stowing was to secure the end to the ring on the bulkhead. This was a safeguard against losing your anchor and all the cable some dark night when you had miscounted the number of shackles passing over the gypsy.

Of course it has been known to happen when the bulkhead lashing had come adrift for the whole lot to go over the side. As a rule occasions like this did nothing to sweeten the relationship between skipper and crew. In fact, a captain has been known to swear at the sight of the bitter end of his anchor cable complete with a raggy remnant of lashing disappearing down the hause pipe into the oggin.

So we restowed the cable, paying particular attention to the lashing, and reattached the cable to the anchor, making sure that the joining link was secure.

If the anchor locker was the foremost compartment of the ship then the aftermost must surely have been the lazaret. This was a storage place abaft the rudder post and once again reached by a small circular manhole in the after deck.

The day we decided to clean out the lazaret happened to be the day that Pete came back to work after a spell off with a poisoned hand. Pete was passing through that awkward stage in life where he was busy increasing his stature from 5ft.5ins to 6ft.3ins and in consequence could at times be just a trifle awkward. He would kick buckets about the deck, fall over matchsticks carelessly left lying around, pour hot water down his sea boots instead of into a bucket, but what he sometimes lacked in co-ordination he more than made up for in his power in a pulling boat. While I would be struggling on the lee oar, Pete would be playing with the other one. It was almost like a toothpick in his great hand.

All the more reason for being concerned when he eventually had to cry off with a septic hand which threatened to turn to blood poisoning. However, when the worst was over he returned to work with a heavily bandaged hand and strict instructions from the doc. not to get it wet or dirty. This state of affairs somewhat bemused Captain Jack. How could he employ Peter's talents about the deck and at the same time guarantee not to get his hand wet or dirty.

Then he had a brainwave! The lazaret! He'd been meaning to clean it out and give it a coat of bitumastic for ages. Two lads down in the darkness scrabbling about the place amongst all the muck and one up top in the sunshine keeping a look out and passing various items up and down to them. Pete grinned at the prospect. Geoff and I were not so sure. Still, it was better than washing paintwork. Anything was better than washing paintwork.

Geoff and I went down for'ard to change into the filthiest gear we had, usually kept in a festering heap in the corner for specially dirty jobs such as bunkering and trimming coal. We emerged minutes later looking like a pair of the raggiest scarecrows ever. Pete, still grinning his head off, jumped to one side as we approached, his beautiful white bandaged hand shielded from possible contamination.

'Suffering Snakes – you smell awful.' He pulled a face in feigned distaste. Geoff lunged at him as he went past.

'Come on, get that manhole cover off and stop messing about.'

Down we went into the evil-smelling, dank darkness with buckets and cloths. We began by throwing up bits of old, rusty wire rope. Rusty old shackles, broom heads, a couple of old duck lamps, the remnants of a copy of the *Christian Science Weekly* and even more unlikely articles. Having finished chucking out through the manhole all the rubbish we could find in the dark, we began mopping up all the water which was slopping around. Bucket after bucket was hauled up by Peter on deck and gingerly emptied over the side before the skipper arrived to poke his head down the 'ole, flashed his torch around and declared himself satisfied.

'OK, lads, that's fine, come up and I'll give you the black bitumastic. I want it all giving a good coat, brushed well in.'

We all trooped after the skipper down below to his paint locker which was next door to his cabin. He raked around until he found a couple of old empty paint tins and two very old brushes. He put these to one side and then pulled out a brand new, unopened five gallon tin of black bilge paint.

'Now be careful with this stuff; if you get it on yourselves it burns like creosote,' he warned. 'Just put enough into each paint tin and pass it down to these two,' he said to Peter, 'and don't be dribbling any on my best teak decks. Be careful how you pour it, this tin should just about do the job – OK?'

'OK, Skip,' we all chorused and set off back to the black 'ole.

'Have you smelt this stuff, we'll be gassed down there before we can finish this job,' said Geoff.

'Oh, you'll be all right,' Pete said. 'I'll stand by to pull you out if you collapse.'

Still grumbling, Geoff and I climbed down into the blackness while Peter spread some old newspapers on the deck and prepared to pour the smelly, black liquid into the two paint pots.

We had to start right up in the farthest corner away from each other and work our way down and backwards and towards the short iron ladder which led up to the deck. It was filthy work; you couldn't see where you'd been, paint ran down the handle of the brush and try as you might you couldn't avoid it running onto your hands. The skipper was right, it stung like hell, but there was no way you could prevent getting it on yourself. After a while you gave up trying and just put up with the discomfort in order to get on with the job. Lying on our sides in a confined space in almost total blackness we concentrated on trying to keep it off our faces. Oh, if only a ship would turn up so we could knock off and have a breather. The tarry fumes made your eyes smart and your nose run like a tap. Your throat became drier and drier and you began to itch all over. It didn't help matters thinking about Peter sitting up there three feet above your head in the fresh air and sunshine. Slap it on, rub it in, wipe your nose on your sleeve, don't kick the tin over.

'How you doing, Geoff?'

'Terrible, can't see a thing.'

'Hope Pete's all right up there, must be terrible to have a bad hand!'

'How's the paint going, Pete?' – This in a louder tone.

'Just under half a tin left – haven't you finished yet?'

Just then we heard the familiar sounds of a ship's whistle blowing one long and three short.

Thank the Lord, our prayers were answered, that was for us, we could knock off for half an hour while we took the pilot off an outgoing ship.

'All right, lads,' a Pilot poked his head down the 'ole. 'Matty and I will attend this ship for you if you want to get done.'

'Thanks a lot, sir,' and then, 'Blast and bugger it,' we both cursed our luck at being on with a watch of helpful Pilots.

BLIMEY! HE'S KICKED THE BUCKET!

No. 4 Watch

We heard the motor boat engine start up and then with the peculiar chuffing noise a propeller makes under water, it receded into the distance.

Slap, dab, pause to wipe your nose, slap, dab.

'Are you still up there, Pete?'

'Give us some more, this can's about empty.'

We heard a thump as Pete jumped down on the deck off the towing rail where he'd been sitting and then CRASH!

'Blimey! He's kicked the bucket,' yelled Geoff as he poked his head up into the daylight.

All over the skipper's gleaming, holy-stoned decks was the best part of a gallon of thick, black bituminous goo.

Pete stood there horror-struck with the enormity of the catastrophe, then with a cry of 'Bluddy-hell, he'll crucify me,' he began scooping up the seeping mess with both hands and slopping it back into the tin. All thought of hygiene, cleanliness or blood poisoning was cast aside as the bandaged hand soaked up the paint like blotting paper.

'Give us a hand here, you pair of molls,' Peter gasped as both our heads popped out of the manhole, and we began to laugh as we blinked owlishly in the sunlight.

Pete was more concerned about what the skipper would do to him than a little thing like a poisoned hand.

'Now look what you've done,' chirped Geoff. 'If you go on like this we won't have enough to finish the job and the skipper won't half get mad if he knows you're wasting it.'

'Come on up and give us a hand,' wailed the luckless Pete, 'before it spreads any further.'

He grabbed the sodden newspaper and threw it over the side but there was still a big, black puddle seeping into the deck. We mopped it up as best we could with our rags and the more expendable items of our clothing.

'I think we'd better mix some Atlas and try and get it off with that,' said Geoff.

'Better still, we'll slosh some motor boat petrol on first and then try Atlas afterwards.'

'What if the skipper comes while we're doing it?'

'Oh, he won't – he'll be having his nap.'

'OK, we'll try the petrol first but be careful, if Matty or anyone else comes down aft smoking we'll all go out in a blaze of glory.'

So we sloshed the motor boat petrol out of the two gallon cans onto the blackened deck. After a while we got the thick of it off with deck scrubbers. Then we poured on the Atlas, a deck bleach which is supposed to be well watered down. The stench of petrol and creosote was almost overpowering but miraculously nobody came aft to investigate the furtive goings on. Anyone carelessly throwing away a fag end would have got more than he bargained for; we would all have been barbecued.

Eventually we managed to get rid of most of the stain, much to Peter's relief. By now he was blacker than either Geoff or me and his face splattered with black was becoming redder and redder with his exertions. We disguised the last traces of the disaster with several buckets of sea water and some judiciously placed smears of sand, a pudding fender and a relocated coil of mooring rope.

'With any luck, Jack won't notice. If he asks why we washed down, tell him the fireman's been feeding the seagulls again and the decks got covered.'

'He won't wear that,' said Pete.

'Well, are you going to tell him the truth?'

'Not unless I have to.'

'Well then, if he doesn't say anything, keep quiet about it and act normal.'

Half an hour later Captain Jack arrived on deck, fresh from his siesta, to make a pot of tea. Leaving his pint pot in the galley he strolled down aft in the late afternoon sunshine, sniffing the air and eyeing the rearranged after deck with some suspicion. However, his only remarks were addressed to Peter.

'Good God, man, I give you a simple job to keep you clean and you finish up looking like a tar barrel. I swear some people can get black using white enamel. Look at the state of your bad hand.'

'Sorry, skip, but I forgot all about the bandage, anyway, didn't they use hot pitch in your day to cauterize wounds?'

Jack grinned.

'Don't be cheeky and change that bandage before you go home or your mother will play hell with me – isn't there a strong smell of petrol around here?'

CHAPTER 14

Determination was the middle name of some Pilots. If they got an inkling of a ship in the vicinity they were as restless as foxhounds after a scent. I think this must have been a throw-back to the days when each man worked for himself and whatever he earned, he kept. If this was the case we had one or two ever-zealous seekers whose dogged tenacity tended to make our lives a lot harder than they might be. Instead of letting the ships come to us they had to go seeking them.

Nowadays with the advent of radar, VHF radio, Decca Navigator and other navigational aids of various sophistication, ships do tend to know where they are. But thirty years ago most ships were pre-war or wartime built and were lucky if they had a giro-compass and an echo sounder. Ships coming from Scandinavia often set a course for Flamborough Head, whose high ground and surrounding cliffs and lighthouse made an ideal landfall. Having picked up Flamborough they simply turned north and steamed the fifty odd miles up the coast until they picked up the Tees Fairway. This was in fact the theory, but it didn't always work out. Sometimes, for reasons of bad weather, poor visibility, darkness, strong tidal sets, wind or a thousand other reasons, they missed Flamborough. They were now faced with the dilemma: having run their speed and distance, did they keep going until they picked up the land, albeit the Humber estuary, coast of Lincolnshire, or even as far south as Norfolk or did they turn to starboard and hope to pick up Whitby Highlight or maybe the loom in the night sky of slag tipping at Skinningrove or even the Tees itself. It is easy to see how a low powered steamer with very few navigational aids, perhaps only a suspect magnetic compass (with a cargo of iron ore), can get into a situation of considerable uncertainty.

The ships on a regular run from Oxelsund or Narvik knew the coast and mostly made perfect landfalls, but a stranger to the port or even a stranger to the North Sea could find himself in a position open to many doubts. This deliberate 'aiming-to-miss' technique is a most useful one.

By deliberately steering a course for a landmark say fifty miles along the coast from your intended destination, particularly if you have made a fairly long passage across open sea and are approaching land at roughly right-angles, when you pick up the land you know which way to turn to follow the coast line. If you have steered to the north deliberately you turn south and vice versa.

The Tees estuary lies in a bay, bounded on the northern side by Hartlepool with its prominent Heugh lighthouse and on the southern side by the cliffs south-east of Saltburn. Regular traders, approaching at night, could often pick up the rose-red glow in the sky a hundred miles away of the tipping of red hot slag, waste from the steel furnaces. This lighting up of the sky was a dead give-away for approaching the Tees for those who knew what to look for, but of course it didn't happen all the time. They were just as likely to tip during the day as they were at night. It all depended on when they tapped the furnaces.

So it happened from time to time a stranger to the port, possibly an over cautious deep sea man, who, as the saying goes, needs a lavatory every time he sights land (as opposed to the coasting man who needs a lavvy every time he loses sight of land!) would be dodging about miles and miles off the land, hoping to make some positive identification to enable him to lay off a course to bring him safely into Tees Bay to pick up his Pilot.

Such was the case one pitch black night as we rolled uneasily at anchor on our station. It was a real classic, dirty night, fresh northerly wind, driving rain, a nasty, choppy sea and black, black as the inside of a cow.

Old Harry, the Pilot on turn, had been roaming the deck most of the night with his little tea-cosy hat on. He was a chunky, powerful little man, about 5ft.2 or 3ins with strong, yellow teeth and big powerful hands. For a little man he had hands so strong it was said as a young man

Cutter Confusion

SEEKING GREEKS

he could pull out a two-tuck splice without effort. His fingers were short and thick without any taper and squared off at the ends. Wandering the deck, muttering to himself, fumbling in the pocket of his much darned old brown cardigan for matches to light his blackened old briar, peering out into the blackness of the horizonless sea, squinting against the rain from the shelter of the overhanging boat deck. Shuffling back into the warmth of the saloon to fetch the ancient, brass night glass. Had he seen something? A dim light, perhaps a ship going north on the collier track to the Tyne? He cursed under his breath as he struggled to focus the unfocusable. The light, barely discernable, hadn't moved.

'I think it's that Greek we have due.' Harry turned to me. 'Have a look, son, see what you make of it.' He handed me the glasses. A dim light danced in the pool of blackness that was the darkened lenses.

'It's just a light,' said I. 'Could be anything, maybe a fisherman.'

'Not a fisherman, too steady for that. It's a ship's stern light about four, maybe five miles and he's just laid there. It's that Greek.'

'Well, if it's just laid there,' says I, 'shall I call him up and get him to come to us?'

'Aye, give him a flash, but first get your mates up and stand by to lower the motor boat.'

The boat was lowered and while I flashed away on the Aldis, Peter tried to start the engine.

Clackety clack, Clackety clack. The pencil beam shone through the murk, picking up the reflection of a million rain droplets as they raced through the brilliant white shaft of light. I had to keep stopping to refocus the lamp on that dull yellowish point of light.

After a while, old Harry's voice came up from down below on the main deck.

'Never mind, son, he's not going to answer. Come down and see if you can get this boat started. We'll have to run off to him, he's not going to come any nearer.'

Down below in the boat, Peter, straight out of a warm bunk, was swearing at the dead engine.

'Bring a torch with you,' he shouted as the boat reared and crashed in the chop alongside. Grabbing my oilskin and torch I jumped alongside him. The lid of the engine box was open and we both peered in.

'Is it the plugs again?'

'It might be, pass the plug spanner and we'll see if they're wet.'

Out came the plugs. They seemed OK. Check the fuel tank – full. Take the top off the float chamber – empty!

'We must have a blocked fuel pipe,' said Pete. 'Pour some juice into the float chamber out of the spare can.'

I slopped fuel into the open top of the float chamber, spilling most of it into the bilges. Peter swung the starting handle again. On the second swing the engine roared into life, black smoke pouring from the exhaust.

'That's it, then. A blocked fuel pipe. That's us well knackered. We haven't time to take it off and clean it out, we've no spares, old Harry is champing at the bit to go off to this ship and that's all of five miles off to the nor'ard.'

'Why not keep pouring petrol into the float chamber; as long as we keep the engine going it'll be OK.'

'Crikey, man, you'll waste more than you use,' said Pete.

'No, I'll get an old bean tin out of the galley and pour it in that – it's worth a try.'

An old baked bean tin was raked out of the ash can, cleaned out and bent to form a spout.

Old Harry, his coat buttoned up around his neck, haversack slung bandolier-fashion across his chest and uniform cap crammed tight down over his bald pate, jumped into the stern sheets with Peter.

'Would you mind not smoking, Mr Blumer,' said Peter.

When we explained why, old Harry thought the request a reasonable one. His knowledge of the internal combustion engine was strictly limited, having been brought up in sailing cobles. I went aft to the petrol locker on the cutter and grabbed a couple of full cans and an extra one for luck, slung them in the stern of the boat and cast off.

As soon as we had let go the painter we swung the big canvas hood up and I began the task of pouring neat petrol into the carburettor float chamber with my right hand while holding a torch with my left hand and keeping the lid of the engine box open with my head and shoulders.

The engine roared away merrily, seemingly enjoying its diet of almost neat petrol. Black smoke and unburnt fuel belched out of the exhaust away astern of us as we crashed and bounced through the blackness.

We seemed to have been running for hours and I had just about used a full can of fuel when I stole a quick look round the edge of the dodger.

'Where's the ship?' I asked Peter as he stood there, his face and woolly hat sopping wet from the combined effects of driving rain and flying spray.

'I think the bluddy thing is steaming away from us, he's no nearer than when we started. Get the boat compass out so I can steer a course, we'll soon lose the lights of the cutter astern of us.'

The engine started to cough and splutter; quickly I sloshed some more fuel in the required spot and then went raking round to find the battered old compass we always kept in the boat. By now, apart from the tantalising dim light we were following, we were completely surrounded by an inky blackness; you felt you could almost reach out and touch it. Not a star, not a glimmer of shore lights, not another ship in sight. We could have been in the middle of the Atlantic a thousand miles from the nearest land – with a dicky engine and two more cans of petrol!

I consciously tried pouring less into that open mouth, trying to make every drop count. As the boat heaved and lurched I tried to steady myself to make sure I hit the spot with my bean can pourer. The effort of bracing myself for a long time made legs and arms ache. Peter and Harry Blumer stood like a pair of fencing posts getting wetter and wetter, not daring to take their eyes off the light we were following in case they lost it in the black murk.

What if we'd been wrong? What if the ship was not for the Tees? We might be following a ship bound for Bremen! It didn't bear thinking about. We had to be right. We would soon reach a point where we would have to turn back or risk being caught miles off in the North Sea at night in bad visibility and no fuel.

Why hadn't I listened to my mother when she wanted me to go into a bank or be an accountant? I'd have been in bed long since, with the prospects of a nice weekend ahead of me.

Instead, here we were chasing after some old Greek or Liberian who didn't even know we were there.

That was the second can empty!

I peeped round the hood again. EUREKA! A red side light. The ship had turned again and was coming towards us. The rain stopped and visibility improved. We were suddenly much nearer to him than we had realised. Flashing our torch at him we drew closer and made out the shape and form of quite a sizeable ship in ballast laid almost stopped.

Passing close under his old fashioned counter-stern we made out his name: *Kate C. Panama.*

Thank the Lord, old Harry's intuition had been right. A pea whistle shrilled out a signal from the bridge and we heard men running along the steel decks. Shouts and a moment's confusion before a cluster of lights was hung over the side illuminating our tiny boat. Some minutes elapsed before a scruffy pilot ladder was flung over the side, the end trailing in the water. A swarthy face appeared over the rail, bathed in the brilliant light of the cluster of bulbs. The ladder was hauled up just clear of the water and made fast.

'OK, son, go to him now,' the Pilot spoke to Peter. 'When I get aboard stay under my lee in the light of the cluster so I can see you. I'll have them get a boat rope handy in case you break down or run out of fuel. Thanks.'

With that he swung himself up onto the swaying ladder and climbed up the black rust-streaked side, carefully avoiding the broken rungs.

'How's the petrol situation?' Pete asked as I crouched over the hot roaring engine.

'I've got this drop here and a full can left,' I yelled. 'It should be enough, we've also got a full petrol tank if we're really stuck; we can always break the copper pipe and milk the tank into the can.'

So we ran back alongside *Kate C.* as the Pilot steamed her slowly in towards the Tees Fairway. After about forty-five minutes we could make out the lights of the cutter at anchor and half an hour later were coming back alongside.

'Where the hell have you been?' shouted the fireman as he threw us the boat rope. 'I've been worried sick about you and just going to call the skipper.'

'You might well ask,' said Pete, 'what's the time?'

'Nearly half past three.'

'Blimey, we've been gone nearly three hours; is the kettle boiling?'

The *Kate C.* slipped past us in the darkness and entered the river as we lifted the boat out of the water and pulled the plug out to drain her. Passing out the empty petrol tins we counted up the fuel we'd used. Four full cans and a bit. Nearly nine gallons!

'The first thing we'd better do is clear that fuel pipe before we need to use the boat again.' Peter went to look for a spanner whilst I stacked the empty cans back in the fuel locker.

At that, old Blood appeared on deck, his face crumpled and unshaven. He tottered along the deck for his ceremonious pee in the leeward scupper. His eyes fell upon the transom stern of the motor boat, hoisted out of the water, and almost level with his face.

'What's up with this 'ere boat? The arse end is covered in soot.'

'Ah well, we've had a bit of bother with the petrol pipe on the engine, Skip, and I think it must have been running a bit rich,' I volunteered.

'You should have come and told me, that's what I'm here for,' said old Blood.

'Well, we didn't like to disturb you, Skip. Anyway, we managed OK.'

'Hell, son, you wouldn't disturb me, I've just been resting my eyes, you know I never sleep a wink when you lads are off in the boat!'

CHAPTER 15

A great many ships using the Port of Middlesbrough were old coal-burning steamers. They were all survivors of the war years and a great many of our regular runners were Swedish, carrying iron ore from Oxelsund and Narvik. These ships were, of course, neutral during the hostilities and so, apart from the occasional mining, were not involved in the war at sea.

Firemen, on coal burning ships, are a very special breed. Theirs is the job to feed the furnaces and keep the steam pressure to run the engines. All the coal has to be shovelled by hand and also, every day each separate furnace fire has to be cleaned with rake and slice to remove cinder and clinker which clogs the fire bars. These impurities have to be pulled out of the fires onto the steel decks of the stokehold in great, red hot heaps, doused until cool enough to handle, and then hauled up in iron buckets to deck level and finally dumped over the side. It is heavy manual work in hot and dusty conditions. Of course, with the coming of oil-fired boilers, this job became simpler and cleaner.

Old fashioned coal burners are now practically extinct and with them the coal shovelling fireman. Not so in the years about which this book is written. Then they were very much large as life and often twice as 'orrible.

They were often a coarse, ebullient lot, usually at odds with the 'crowd' on deck (i.e. the sailors). Often they were hard drinkers, womanisers and a constant pain in the neck to Masters who strove to keep order on board when at sea and fend off the police and Customs' Black Gang when in port. A favourite spot to hide contraband bottles and cigarettes was amongst the tons of bunker coal – not very subtle but requiring hours of back-breaking digging to find!

All the tugs on the river, with the exception of one, were coal fired as were many of the little coasters which ran to and from the Continent with scrap, slag, fertilizers and other bulk cargoes. The coal hoists were kept quite busy bunkering these ships and coal from the nearby Durham coal field was still a cheap and plentiful fuel.

Most of our firemen, employed on the steam cutter, were not deep sea firemen as such. The odd one or two had been in coastwise ships or stokers in the Royal Navy but, by and large, they were sent down by the Labour Exchange, having declared a knowledge of firing marine boilers. Some were very good, some were eager but inexperienced, others were a washout. We soon found out which were which. Some stayed for years and we became very pally with them, others stayed a few watches and were never seen again. You could usually tell when they arrived for a two-day spell with a loaf of bread and six bottles of Guinness that they hadn't any real idea. When the Guinness ran out after about twelve hours, they simply walked ashore at the first opportunity to find the nearest pub.

If the fireman was a friendly type, the cutter lads often had a good relationship with him. He was someone to talk to during the long, tedious nights when everything was quiet.

Jack Silverton would always be handy to throw you a boat rope or grab your arm as you leapt for the cutter's deck from a pitching boat. In return we would sometimes help him dump his ashes in the early hours. He was a stocky, squat, tremendously powerful man, with the torso and arms of an old artillery man. His clean-shaven jaw was one any pugilist would have been proud of as it was set firm against life's tribulations. He was blessed with twelve children, six boys and six girls, so I suppose he was used to tribulations.

He always shaved with bar soap and an open razor and rolled his own cigarettes. He only once asked me to cut his hair for him and took the consequences with good humour, his eyes twinkling behind thick lensed spectacles.

'Never mind, son, I would have made a bigger mess if I'd done it meself,' he chuckled, and ambled back into the stokehold.

Jack's relief on the other watch was a totally different type. Much younger and a bachelor, he told us tales of his female conquests, real and imaginary, I suspect.

Biff took great delight in pulling our legs about the ridiculously low pay we received. He would clop, clop along the deck wearing enormous boots without laces and an immaculate boiler suit. He would call us 'Cutter Company hand rags' or 'Strickland's slave labour'. He was quite right, of course; maybe that's why he nearly always passed these remarks within ear shot of a member of the Cutter Company Committee. It never did us any good but I suspect his remarks were aimed not at us but at others by whom we were exploited.

The galley was our common meeting ground and we ragged each other mercilously at times. Biff had been in the Royal Navy for a time and so had seen quite a bit of sea-going life. His naval discipline showed in his neat and tidy appearance and his dependability. No matter how much ale he had supped the night before, he was always on the jetty at eight the next morning to relieve Jack. That was more than could be said for some!

When Biff eventually left we had a succession of replacements. Some, like Biff's own brother, were with us for some time and proved quite dependable types. One such character was Cliff Larvin, again an old Naval man. He seemed to spend his off duty time amongst the Redcar fishing fraternity while they were engaged in the summer occupation of taking day-trippers 'round the buoy' for half-a-crown. Cliff was a bit of a lad for the girls, a flashy dresser in drape suits and crepe soled suedes. He walked with the peculiar rolling gait and soft-footedness of a slightly tipsy pussy cat. He was clearly unhappy at working such long hours at a stretch and would often gaze across the black waters of the bay at night at the beckoning shore lights of Redcar's Promenade – his happy hunting ground – and sigh,

'What am I doing here with all those gorgeous females just waiting for me over there?'

Eventually, the lure of the neon lights was too great and he left for a

FACE EACH OTHER ON THE BARE BOARDS OF THE HALLWAY

day job – with the Council, I think.

In Cliff's place came a great character, Steve Valley, one of the most amiable rogues you could ever wish to meet on a day's march. Steve was a fireman *par excellence*. His skills in the stokehold were akin to a *chef de cuisine* in the kitchen. I swear that when, to all intents and purposes, his fires were almost out, reduced to glowing embers on the grates, he would go down below and, at the rattle of his shovel, not only would we have a head of steam but the safety valve would blow off! He often did this deliberately to panic the engineer. He worked with 'The Gook' for a while and took a delight in tormenting him.

Steve's big weakness was booze. If he went on the binge he would be missing for days. His wife and family never saw him until, maybe, the police brought him home after being engaged in some brawl. He would be a man of about fifty-five, but still couldn't resist a good scrap on a Saturday night. Bald, he had one cauliflower ear and a nose that had been broken so many times it was like a piece of India rubber. When he leered at you he had three or four teeth hanging like pickle stabbers and the rest was just gums. Once asked why he didn't get some false teeth on the National Health, he replied that he had had some once but one Saturday night coming back over the Transporter Bridge, having had too much in the Port Clarence Hotel, he was sick over the side of the car in mid-stream.

'I haven't lost me teeth. I know where they are but I can't swim!' he cackled.

For some reason or other Mr Strickland, our Superintendent, had a soft spot for him. If he failed to turn out and booze was suspected, Strickland would call round at his house. Steve would drag himself out of bed to answer the door and the two would face each other on the bare boards of the hallway, Strickland, immaculate in pin-striped suit and homburg hat; Steve, in stained and dirty old 'combs', his feet so black that the Super thought, at first, he was wearing black socks!

'Pull yourself together, man,' thundered the Super.

Steve would pick up his overalls from where he had thrown them on the floor, pull on a pair of old, worn out boots – no socks – and follow him meekly out to the Jaguar, which, by now, had attracted a noisy mob of urchins. The same urchins that had cheered and shouted as Steve fought like an alley cat on his own front grass with his two eldest boys as they quarrelled after a night 'up the club'.

Once back on board the cutter, Steve would solemnly preach to us about the evils of booze and the sheer futility of fighting, twisting his nose and ear into grotesque shapes to emphasise his point. Steve was his own worst enemy. His long-suffering wife used to appear at the Pilot Office on pay days as Strickland would only give Steve his money on alternate pay days. That way the family did get something to eat. Otherwise he just never went home until it had all gone in the nearest pub.

If ever a fireman had the soul of a poet, Billy Bush was the man. He was an incurable romantic as far as the sea was concerned and badly wanted to be regarded by all as an 'Old Salt'. In fact, his background was about as far away from the sea as it is possible to get. He had been a regular soldier serving in the Cavalry in India for many years – maybe all those spectacular Indian sunsets had had a lasting affect on him. Certainly Indian service had given him a taste for curry and garlic which was something we all regretted!

He was a tough, wiry scarecrow of a man, very ginger, with deep sunken eyes, an even deeper vibrant voice and an enormous 'Adam's apple'. Much later in life he grew a dense ginger beard and took to wearing nautical clothes.

Whenever he got you in a corner and began one of his long sagas of the sea, you received the full force of 'the spices of the East'. Working as fireman with us was, he pointed out, just a stepping stone to greater things in the nautical field. He had every intention of becoming, as he put it, 'a blue water fireman'. What exactly that meant wasn't very clear, except it involved flying fishes and dusky people of tropical islands.

As a fireman, Bill was adequate; as an orator and reciter of tales, he was much better. As soon as the opportunity arose he was off. He'd heard of a berth going on one of the steam 'flat irons' of the Gaslight and Coke Company trading to London and, raving optimist that he was, he jumped at the chance. Someone omitted to tell him that, when the ship sailed late at night, he was one of only two firemen signed on for the thirty-six hour run to the Thames, only the other fireman hadn't turned up. Poor Billy, by the time he arrived in London after four days of bad weather, he was knackered.

I think the experience must have cured him once and for all of ever wanting to go a-roving – bugger the flying fish! He found his niche with the Tees Conservancy Commission as reserve lighthouse keeper and served as such until he retired. Very definitely a character.

Last but not least we had a diminutive bundle of energy in the person of Jeff Woodcock. Jeff had little or no experience in firing boilers when he arrived, but by sheer bubbling enthusiasm he learnt fast and, apart from one or two mishaps, proved a great little guy. The only thing that put a damper on Jeff was bad weather. He hated it. It wasn't so much that he was seasick, and he was very seasick, but I think he was just plain scared. Seeing us working in the boats in snotty weather, he would often remark that there wasn't enough money in the mint to tempt him into that boat.

'I don't know how you fellows put up with it,' was his favourite expression.

But put up with it we did and, in fact, quite enjoyed a lot of it. Mind you, we shared Jeff's dislike of bad weather, but for different reasons.

CHAPTER 16

Although our official title, in the Pilotage Service, was Boathand Apprentice, our predominant training was undoubtedly boat work. We were encouraged and expected to spend a certain amount of our off-duty time accompanying Senior Pilots during the Pilotage of small ships. This gave us an insight into the arts and skills of ship handling as practised by the more experienced Pilots.

Some of the lads had fathers or uncles who were Pilots and came from families with a long-standing tradition in the Pilotage Service. Others, like myself, were completely new to the job and had to learn fast.

It was probably something to do with the fact that we were 'the lads' that most of the older Pilots treated us with a kind of amused tolerance. Many of them did not take us too seriously, but it gave us a chance to be with one character at a time rather than have to deal with a whole watch. Some of these older men were very near retirement age and indeed the odd one or two were past retirement but for one reason or another had stayed on. We thought, specifically to torment us.

It was once said jokingly that what a Pilot lacked in intellect he more than made up for in low cunning!

To a certain extent this was true. They were brought up in a hard school; to survive was to succeed. The successful ones had survived a far harder apprenticeship than we were serving and this and the ensuing time had left its mark on many. They had an inbuilt sense of independence and self-reliance which often brought out the ultimate character of each man. Some were quiet and introverted almost to the point of being morose. Others were full of humour, almost flippant, and between the two extremes were all shades of grey. No two characters were alike.

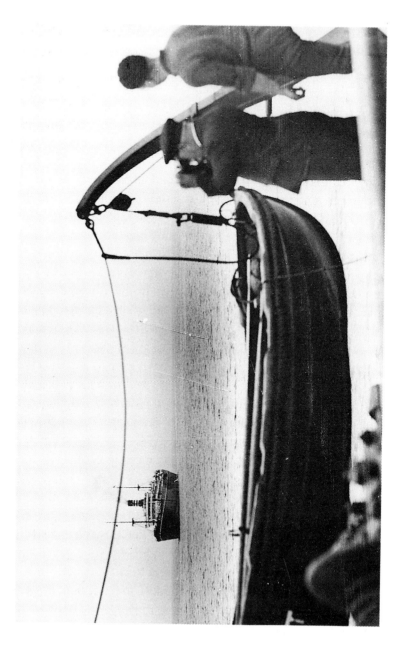

Oriental visitor

Probably the most flamboyant of the elderly men was a crusty, old bachelor affectionately known by all his colleagues as Stanley. He was delightfully eccentric with an impish sense of humour which was not always appreciated. He often arrived on duty in an orange shirt and wearing an immaculate camel hair coat over his uniform jacket. To see him off duty in the town reminded you of a show biz character filling in time before the evening performance at the Empire Theatre. He drove a bright red SS Jaguar open sports car as immaculately turned out as himself, when those that could afford cars pottered around in Austin Sevens and Morris Eights. Later he had the first XK120 in the town, a beautiful ivory colour, which he drove at funereal pace up and down Linthorpe Road wearing a chequered county cap and yellow driving gloves. Everyone knew Stanley, he had style.

He also had a tremendous pride in his profession. When faced with a tatty little Dutch clog (small coaster) he would sometimes throw a little tantrum, leaving no one in any doubt as to his disgust at having to waste his talents on so humble a vessel. I suspect that once on board he was quite capable of conveying his displeasure to the hapless skipper and his crew and would probably go into one of his black moods.

In direct contrast he would positively sparkle with giggling good humour if there happened to be a passenger-cargo ship due and he looked like being on turn for it. I think P and O Liners were his favourites. He would stand in the forepart of the boat as we ran towards the *Canton* or the *Somali* and with an approving shake of the fist would declare in his strange, impedimented speech.

'B-B-By Qui-Quiste, son, that's a man's ship, look you,' and dissolve into a spasm of delighted chuckles and giggles.

On reaching the ship's side he would stand on top of the engine box, draw himself to his full 5ft.5ins, cup his hands to his mouth and address himself to the Master on the bridge wing.

'Bring her round to port, sir, I'm coming on board.'

One always got the impression that Stanley followed his profession mainly to keep himself amused. He didn't need the money, he was always dressed impeccably, though sometimes a trifle oddly. His clothes were expensive and his pipe always had that whiff of Balkan Sobranie. It was said that his father, an old sailing ship skipper, known as Ping Pong, was even more eccentric, but I was not familiar with that gentleman.

However, it just so happened that one afternoon we were working with Stanley's watch. The cutter was at anchor on station at the Fairway Buoy and Stanley was on turn for a big BI for the LNER Dock. The ship had given an ETA the day before on leaving London and it was imperative that it improved on it or else it would miss the tide. Stanley paced the deck muttering to himself, pausing only to relight his pipe. The sea was calm and the visibility was very clear but the sky overcast. The time ticked by as the tide reached high water and began to ebb.

We scoured the horizon to the south with our ancient glasses and eventually spotted a blob of something miles away. Could it be the BI?

Stanley started to fidget even more, striking matches and blowing great clouds of pungent tobacco smoke around the decks. As time went by it became abundantly clear that the blob was not the BI – he would be charging up at a good 17 knots. This blob was floating north punching the tide at about four and a half.

'What's that to the south, son?' Stanley asked as he spotted the blob getting nearer.

'Don't know – something very small, certainly not the BI; what else have we got due?'

I dashed down into the saloon to the drawer where all the orders were kept. Apart from the *Nyassa*, the BI, there was just one more card, a little converted Thames sailing barge called *The Lady Sonia* that had been due for the last three days. She was for Stockton! That's as far as you can go on the Tees.

When I told Stanley I thought he was going to swallow his pipe. His face went a puce colour.

'Quiste, son, where's the BI? I'm expecting my tea on the bridge from a silver tea service – look you – and you tell me it's a bloody BARGE!!'

But the barge it surely was. Chugging up from the south with a single mast like a long broom handle on the foredeck, one hatch and a wheelhouse for all the world like a shoe box stuck on right aft almost as an afterthought. These craft – there were two or three of them that used to visit us from time to time – were so low in the water that you actually had to step down from the motor boat onto their deck!

Of course, to make matters worse, down to the south was a smudge which quickly became the BI, thundering up – too late, he had missed his tide, the dock gates would be shut.

Stanley was so mad he hopped and skipped with sheer frustration. He bit clean through his pipe stem and showered himself with hot ash.

Then a strange thing happened. The little barge began turning off to starboard. As we stood and watched it, it described a great wide circle, taking almost ten minutes to complete. So slowly did it turn, we wondered if its steering had broken down or maybe one of its engines had packed up and it had become uncontrollable. Some of these little craft had been fitted with marinised lorry engines which roared away in minute engine rooms. But having completed an almost perfect circle the little ship steadied on its helm and set course direct for us.

Now, from time to time, if the opportunity arose and we were nearing the end of our watch, we would persuade the skipper to let one of us go on board an inward bound ship with the Pilot to do the passage with him. It occurred to me that this might be one such opportunity. The ship was for Stockton, which was the maximum distance the river was navigable and also it was only an hour to go to relief time. So permission was asked and after much persuasion was granted. I rushed to get a wash and brush-up and generally make myself presentable, after asking Stanley's permission if I could accompany him.

By this time he had simmered down sufficiently for me to approach him without chancing having my ears chewed off. We set off in the motor boat to board *The Lady Sonia*, Stanley standing deliberately with his back to the little ship as he stood and watched his beloved, majestic, BI come, elegantly, to anchor about two miles away. As he turned to face the little ship an expression of obvious distaste crossed his face.

To me, it was all a great adventure; the fact that I would be hours late home that night didn't bother me. Punching a spring ebb tide to Stockton

132 Cutter Confusion

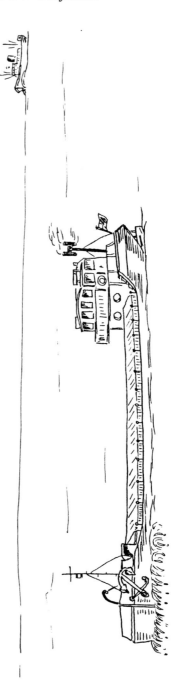

FIVE DAYS FROM SELBY

in a six knot ship was not the quickest way home!

We swung alongside and throttled back to match his speed. I followed Stanley along the tiny, narrow, side decking to the wheelhouse which looked as though it had been concocted out of hardboard and plywood. Great black greasy steering chains emerged from underneath the wheelhouse, led round rollers and were attached directly to an iron yoke on top of the wooden rudder.

The door opened and a ruddy-faced man in a much darned, navy blue gansey beckoned us to enter.

'Afternoon, Pilot,' he grinned, swinging half a turn on the enormous ship's wheel which completely filled the rear part of the wheelhouse. 'Sorry I'm a bit late.'

'Where are you from?' asked Stanley. An academic question as I don't suppose he gave a damn where he was from.

'Selby,' grinned the skipper.

'No, I don't m-m-mean you p-per-personally, I m-mer-m-mean what was your last port?'

'That's right, Pilot – Selby. It's taken us three high waters and a couple of heavy dews but we left Selby five days ago.'

'G-Goo-Goo-Good God,' stuttered Stanley. And then after a long pause he asked, 'What did you change your m-mer-m-mind for, der-der-down to the south when you turned her right round?'

'Well, you see, Pilot, me and Bill, that's the mate, are the only two on board and seeing as we're working watch and watch, I thought I wouldn't get Bill up – so I lashed the wheel wi' a bit o' starboard 'elm on and went down for a shave and a wash.'

'Goo-goo-good God,' stuttered Stanley again.

'Talking about watch and watch, Pilot,' continued the skipper, 'Bill's been bad in his bunk for the last two days so I've been working Chinese watches, and I'm about knackered. I wonder if you'd mind taking the wheel for a bit.'

Stanley reacted as though he'd been stung. He eyed the fearsome object referred to, lurking in the back of the wheelhouse with its great, greasy black chains turning on a black iron spindle like some ancient instrument of torture, it must have been all of seven feet high, and declined the invitation.

'You've been doing a ger-ger-good job for the last two days, another couple of hours won't kill you.'

I made myself as inconspicuous as is possible in a wheelhouse the size of a decent wardrobe and said nowt.

The engines roared away under our feet.

CHAPTER 17

If there was one word which was guaranteed to produce reaction from all sections of the Pilotage Service it was – bunkers!
The Pilots disliked it because it meant upset to their routine and often they had to make do without their usual creature comforts. The skippers and engineers disliked it because it often meant much-needed rest being lost and a lot of extra effort having to be made, often when they had already been up and about for twenty-four hours. Finally, we, the lads, disliked it because it meant a lot of extra hard and dirty work, often coupled with five or six hours of unwanted overtime.
But the fact remained that the cutter, being a steamer, ran on coal and coal had to be taken on board every ten or twelve days depending on how much steaming time we had put in. Time spent at anchor on station demanded little fuel but when bad weather prevailed it followed that the cutter had to spend more time underway and fuel stocks soon ran low.
The cutter could carry somewhere in the region of forty tons of coal, of which about twenty-six tons would drop straight into the main saddleback bunker but the remaining fourteen tons had to be trimmed by hand into the wing bunkers on either side.
When big ships took bunkers, they employed proper trimmers, men whose job it was to shovel and pack the coal into the more inaccessible places that the coal would not run into itself after being delivered down the chute. They were paid on tonnage delivered and would not work in confined spaces. Unfortunately, our cutter, being rather small, not only took a fraction of the bunkers of a big ship but also its side bunkers were quite small compartments holding only about ten or twelve tons on each side. For this reason, professional trimmers just didn't want to know us.

So it fell upon the engine room staff and the lads to do the trimming for which service we received a shilling a ton shared between five of us! It usually worked out about eight or nine bob each and never was money harder earned. Usually, when the steam cutter left the station to go up to the Dock Cut Coal Conveyor, for bunkers, the diesel cutter, the much smaller *W.R. Lister*, would come down on station for the day. This meant that the duty Pilots would have to move into a brick-built shore station which was a remarkable place only in the complete lack of imagination that had been employed in its design and furnishing. Small wonder that they moaned and groaned on bunkering days.

If the *Lister* was not available for one reason or another we simply left one of the boats with a good supply of petrol in cans and hoped for the best!

We drew lots and one lad would stay back to work the station for the five or six hours it took to bunker, and the rest, including a lad who had to stay back from the previous night watch, would remain on the cutter to do the dirty work. As we steamed up the river we would make ready by removing the tarpaulin and hatch boards from the bunker hatch just abaft the funnel on the boat deck. We had to arrange the hatch boards in a kind of funnel to direct the stream of coal into the bunker. The driver on the conveyor was a notoriously bad shot and often deposited several tons on the side deck before he was able to stop his infernal contraption. This of course made more work for us, so we all heave ho'ed on the guy lines of the great big black chute and made them fast as best we could so it didn't flap around when he started pouring coal.

The manhole covers on the main deck were removed from the side bunkers and all loose items that could be stowed away were removed as we tried to keep the decks as free from clutter as possible. The skipper usually made himself a large pot of tea and then disappeared below, closing all the outside and inside doors as he went as coal dust creeps through every nook and cranny.

We climbed into our dirtiest assortment of rags and worn out, old sea boots, pinched a piece of new mutton cloth to cover our nose and mouth with, grabbed a trimming shovel and stood clear of the black cloud of dust as the first truck load was tipped and began churning its way along the conveyor belt, finally crashing down the enormous rusty iron chute into the black depths of our bunker.

For a few minutes, the noise was appalling. The dust rose, enveloping everything. We cringed in the galley with the door shut until a shout from the shore gang told us that the first truck load had been delivered. We emerged onto the decks – if the aim had been good there was little for us to do as the first truck load would drop straight into the bunker. Another shout from the conveyor would send us scuttling back as the next truck load was on its way. Rattle. Rattle. Crash. Bang. Crash. The great lumps of coal banged and bounced off the edge of the coamings onto the deck, one of the guy ropes had loosened and the chute swung a couple of feet, showering about two or three tons onto the side decks, some bouncing off the gunwale and plopping into the water.

'Stop the bluddy thing,' shouted the engineer, as he saw more and more of his fuel going over the side. After a minute or so the cry was heeded and the machine ground to a halt.

'Right, lads, let's get this lot shifted,' and we began shovelling the pile on deck down the side bunkers. Meanwhile the engineer and fireman went up top to secure the guy lines in another position.

By this time the whole atmosphere was black with glistening dust. It was literally everywhere, even lying on the surface of the water surrounding us. We sweated away in clumsy ill-fitting garments until the bulk of the coal was off the decks and then the Chief declared he was ready for the next wagon to be tipped.

Rumble. Rumble. Rumble. Crash. Crrrash. The coal came hurtling down again. This time we didn't bother taking shelter, we just stood out of the immediate way of the bouncing torrent, we couldn't be any more filthy than we already were, and sweaty to boot! This time, of course, the main bunker soon filled and the coal, still pouring down, cascaded in a torrent on both side decks, port and starboard.

'Stop the bluddy thing.' The engineer, now almost as black as us, danced a little jig and waved his arms at the bloke in the control cab. The flow stopped.

This was where we started earning our coppers!

Two of us slid down the manhole on either side bunker and, in the blackness, after our eyes had become used to the dusty darkness, began trimming the coal out of the main saddleback and into the side bunkers. With a bit of luck, if the coal was reasonably small nuts, we could get it to run with the pointed trimming shovel. Using it like Paul Robeson used his paddle in *Sanders of the River* we could flood the first few tons of coal into the smaller bunker without too much effort, but as the angle of repose got less and less the stuff was less inclined to run and you had to start shovelling hard. It got hotter and hotter in those side bunkers. As we packed the coal up the bulkheads, less and less light came into the bunker until we had to call for a couple of duck lamps to be passed down so we could see where we were working.

We were standing thigh deep in the stuff. Lying on it, sitting in it, breathing it, cursing it. Coal dust choked us, up our nostrils, in our throats, hair, eyes.

We still had several tons to pack up the bulkheads of that bunker.

A cry from the deck, a faint whirring, then an almighty crash as the last truck load was tipped up against the steel bulkhead a yard from your head. Suffering Snakes! How much more? You had to change your grip on the handle of the shovel and put your hand through the hole in the handle so you didn't smash your hand inadvertently against the steel bulkhead in the dark. The top of the handle took the blow instead.

More coal surged into the side bunker under the pressure of that being tipped in the saddleback.

Now we were working in an ever-decreasing space as we packed coal in underneath us and banged our heads against the deck above. It got even hotter. I had only to stand upright and my head and shoulders were in the circle of the manhole. Marvellous fresh air. I blinked in the dusty daylight at the Chief who was leaning against his shovel on deck supping a great pot of steaming tea.

'That's the last, son, how are you getting on?'

'I'm bluddy sick,' said I, 'but thanks for asking.'

'You'll need to get a bit more in those side bunkers before you can get the hatch boards on,' said the Chief, completely ignoring my remark. 'I'd give you a hand but I can't get down the 'ole.'

I ducked back into the bunker and resumed the back-breaking chore.

'What did he say?' said Pete grinning at me from the heap of coal he was attacking in the corner. The whiteness of his eyes and teeth and the red of his lips were a clean colour contrast to the matt black of the rest of him.

'Coo, you don't half look mucky,' I chuckled and spat coal dust into the opposite corner.

'Have you seen yerself, you ain't exactly snow white,' Pete replied.

'He says that's the lot, but we can't get the hatch on yet, so he wants some more packing in 'ere.'

'Christ, if we pull any more in 'ere we'll never get out ourselves.'

'Stop talking and trim coal,' says I.

'We can't get any more in 'ere,' says Pete, 'I'm on my knees now and my head's touching the deck head.'

'OK. I'm sick anyway, we'll have to chuck a few hundredweight down off the top deck and bung it down the 'ole 'til we can get the hatch on.'

We both crawled out on deck and sat on the gunwale getting our breath back, looking for all the world like a couple of raggy old crows. As the folds in our shirts and pants fell out the coal dust dropped like confetti all around us, sparkling in the pale sunlight.

Peter sucked at a bleeding knuckle and I pulled out a ragged handkerchief, showering him with dust. I blew my nose to try and clear my sinuses of the clogging muck.

It was then that a familiar figure appeared, standing on the edge of the coal staithes, his immaculate suit pin-striped and superbly tailored, in sharp contrast to his filthy, ramshackle surroundings. A dark grey homburg topped his sartorial elegance as he picked his way as delicately as a well-groomed tom cat along the quay towards us – it was the Superintendent.

'I have some stores in the boot of my car for you, can you come up and collect them?' He waved an elegantly suede-gloved hand in the direction of a beautiful pale blue Jaguar saloon parked a discreet distance away from all the dust and confusion. Shuffling to our feet we climbed ashore as he carefully crossed the boat deck and disappeared down below to find the skipper.

Peter wiped his hand on the seat of his pants before grasping the polished chrome handle of the boot lid. Inside, about a dozen gallon tins of paint, mostly buff and black. A ball of tarry marlin. A coil of point line. A cluster of shackles and a couple of buckets.

Even the interior of the boot of the car somehow imparted an air of opulence which made us react with a sort of semi-reverence. A Mk.V Jaguar was a chariot of the gods!

We removed the tins and closed the boot lid, then we sauntered round the car shielding our eyes against the reflection as we examined the interior with knowledgeable and appraising looks. Both Pete and I were enthusiastic drivers of our parents' cars but Hillmans and Ford Eights were small stuff compared with three and a half litres of Jaguar muscle.

Carrying the cans of paint four at a time we soon shifted the store.

'Have you seen his SS100?' Peter said as we collected the last of them.

'No! Has he got an SS100 as well?' I replied incredulously.

'Aye, he was at the breakwater with it a couple of Sundays ago.'

'Blimey, some people have all the luck. Look out, here they come.'

Old Blood and the Super. arrived on the boat deck from down below, deep in conversation. On seeing us they stopped and the Super. turned to us, saying,

'I've left some coir fendering in the office; the two of you go and collect it and also the petrol coupons – you might have to break the coil to carry it between you but you're strong enough lads, you'll manage.'

'And don't take all day,' Blood growled, 'We have to go round to Corporation Quay to take water after finishing here and the ebb's away now.'

'Aye, aye, Skip,' we chorused, and scrambled ashore, glad of the opportunity to stretch our legs. There were a couple of glamorous females in the office and we welcomed the opportunity to chat them up, especially when 'El Supremo' was absent and otherwise engaged.

As we walked up Dock Street past the Albion and the Lord Byron, we speculated on our chances of getting one or both of the office girls to go roller skating at the Town Hall that coming Saturday.

'We won't get Wendy talked into going, she'll be going out with Doug,' said Pete, 'T'other one might.'

Wendy was the typist and had already taken a shine to Doug, one of the lads on the other watch working with Alan and Stanley.

'Well, Doug's working this Saturday, so she might come along to make the party up.'

We weighed up our chances as we strolled up into Bridge Street in our filthy bunkering rags and stopped outside the windows which proclaimed in large letters of gold 'PILOT OFFICE'.

Peter produced half of an ancient comb with several teeth missing and we stood outside the massive door preening ourselves in the burnished brass nameplate attached to the wall at a more or less convenient height. Somehow we only managed to make ourselves look more ridiculous with dead straight partings revealing the whiteness of the scalp as a vivid contrast to our smudged and sweaty faces.

I knocked the thick of the coal dust out of the cracked old comb and returned it to Pete. We glanced each way up and down Bridge Street and went in.

The office consisted of one big room and two smaller ones on the ground floor. The big room had a floor to ceiling frosted glass partition with an access door and a small enquiry hatch at desk level. Behind this partition was the general office beyond which was one of the smaller rooms – the sanctum sanctorum – the Super's private office. Outside the glass screen was access to the other small room used as a cloakroom and store with toilets leading off it. A staircase ran up out of the foyer to the boardroom upstairs.

We tapped at the glass enquiry hatch and waited respectfully. The noise of a typewriter stopped and a figure approached. The window opened and both girls smiled at our dishevelled appearance.

'No coal today,' said June.

'We've come for the coupons and some fendering and, what are you doing on Saturday night?' blurted Peter all in one breath.

The two girls looked at each other and then at us and began to giggle.

'Have you seen yourselves?'

Peter and I turned to each other and both our faces slowly contorted into toothy grins.

'We've been looking at each other all day, thought we'd come up here for a change of scenery – what about Saturday night at the Town Hall?'

'You'd never get clean in time,' Wendy chirped from behind her typewriter, 'Besides, Doug's taking me but I might bring my sister if you promise not to make her sea-sick.'

Just then the noise of a car pulling up outside sent us scattering into the back to find the fendering.

'Give us the coupons and we'll be off, here's the gaffa back,' said Pete, as the Super. strode through the door.

'Come on now, you lads, the Captain is waiting to get away.'

Pocketing the petrol coupons we cut the lashings on the roll of fendering and, unrolling it, we flaked it out on the floor of the office until it was in handy lengths to carry over our shoulders, then each picked up one end of it like a huge length of carpet. We marched out in Indian file, winking at the girls as we passed the open hatch. It was bulky but not too heavy and we managed it back to the coal conveyor with a couple of stops to transfer the weight from one shoulder to another.

Passing the Albion again we were almost knocked flying by a couple of burly trimmers, staggering out a bit the worse for wear, having slaked their thirst on seven or eight pints of best Vaux bitter.

'Watch out that thing doesn't strangle you, kidda,' one of them shouted as they both fell about laughing in the road.

One of them, the fatter of the two, grabbed his mate by the back of the enormous leather belt he wore to keep his paunch under some kind of control, and yanked him back on his feet again to continue their tottering way back to a big Swedish steamer taking bunkers at the coal hoist on number ten quay.

I often wondered how they, the professionals so to speak, managed to stay relatively clean while earning their money, whilst we lads managed to get so incredibly filthy. Pausing to let a shunting engine move a flight of coal trucks across the lines which crossed the cobbled road, we got a cheery wave from a tug-boat skipper off home on his bike for his dinner.

'You'd better hurry up, lads – Old Blood's stamping round the after deck like a caged tiger. Said if I saw you to tell you to get a move on. I think he's worried about getting round the point now the ebb's away.'

We shouted thanks and shuffled a bit faster as a token gesture.

The coal staithes we were laid at for bunkers were situated in a man-made cut off the main channel of the river. It was also the entrance to the LNER Dock, hence its name Dock Cut.

The Corporation Quay to which we had to move to fill our tanks with boiler water lay up-stream alongside the Transporter Bridge and to get there we had to steam round the Dock Point through an angle of about 150 degrees. This in itself should have presented no problem but the snag came as you tried this manoeuvre with a vessel as low powered as the cutter. You simply could not get enough 'galley way' (speed) on from a standing start in such a short distance. The Dock Cut was too confined to take a good run at it and get enough rudder effect before you were round the corner, out of the slack water of the Cut and into the full ebb stream of the river. Added to this was the complication of a line of massive iron mooring buoys in a row several hundred feet apart which you had to avoid getting tangled up in.

Old Blood was a worried man, he'd had trouble before. His state of mind was indicated by his constant nattering at us and his cigarette holder clenched between his teeth so it stuck out like a chapel hat peg.

As we made the mooring ropes tidy and readily to hand for Corporation Quay the skipper, in his wheelhouse, backed and filled the old lady until he was satisfied he was in the best possible position to give him a good chance of bringing off the manoeuvre first time. Lining her up in the still water of the Cut he twirled a treble ring FULL AHEAD on the telegraph and lit another Woodbine, fixing it in his long holder. The propeller started to move ahead as the engineer, with a fresh pot of tea in his hand, turned on all the steam he had. The thrust from the wash barely disturbed the water under the counter as the vessel began to move ahead.

In the Harbour Master's office on the Dock Point the watch keeper put down his newspaper and began taking an interest in the proceedings. If anything went wrong he was ready to run for it. The bluff bows were heading straight for him at an ever-increasing speed. He had not forgotten the time when the Assistant Harbour Master had stood right on the knuckle end as the flair on a big ship's bow had almost passed over him, screaming his head off down a megaphone to 'back off'. The mate on the fo'castle of that ship had taken upon himself with great presence of mind to let go the anchor. The said piece of hardware had landed, all five tons of it, with an almighty crash not ten feet away from him, passing through the rotten timber together with about fifteen fathom of cable. At that point the AHM lost all interest and went indoors to change his trousers. So it was with considerable relief that the watch keeper saw the cutter's bow clear the knuckle and poke out into the strong ebb tide.

Old Blood heaved away on the wheel and the steering engine down aft clattered the rudder hard a'port as the bow shaved the woodwork of the knuckle. By now the engines were in full cry and the propeller flailed away with all the thrust and power of an egg whisk.

'She's got to come now,' muttered Blood to himself as he dragged fully half the Woodbine down in a single breath. His knuckles whitened on the wheel as he willed the old girl's bow up into that hostile current. Crabbing her way sideways she swung in an ever-increasing arc wide across the full width of the channel.

BOOOUUNGG!! The noise came like a great Chinese gong as our whaling strip on the stern clouted the mooring buoy a resounding blow. Clouds of pigeons flew out of their roosts in the Dock clock tower and the crew of the *Francis Samuelson*, moored at his own tier of buoys close-by, held their ears as the sound reverberated through the still air.

Old Blood, pleased as punch at his successful manoeuvre, dropped the wheelhouse window and, with a slightly quizzical look, asked Arnold on the foredeck, 'Did we touch that buoy, son?'

We steamed jauntily up river, listing slightly to starboard, to keep our appointment with the water hose on Corporation Quay, leaving a trail of coal dust and a slightly dented buoy behind us. It hadn't been such a bad day after all.

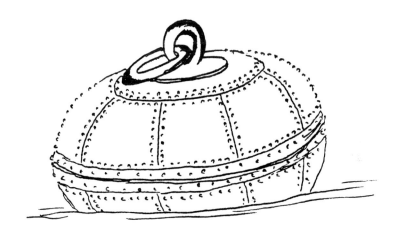

CHAPTER 18

At last summer time arrived with its long days and more importantly its shorter hours of darkness. We made the most of the balmier weather often working all night in shirt sleeves and wearing only sand shoes.

The summer seas took on an altogether softer and less hostile atmosphere. An air of temporary relaxation pervaded our working lives. I think it must have been the arrival of certain microscopic plankton in the water which, at times, caused the sea to glow and sparkle with an eery phosphorescent light. The bow wave thrown up by the motor boat, glowed as if incandescent and each blade of the propeller wash could be clearly seen as it cut successive swathes through the black water astern of us. On moonless nights the velvet sky shone with all the constellations of the northern hemisphere, each reflected in the darkened waters. Shoals of fish, startled by our boat, would scatter in beautiful fans of glittering greens and blues to regroup and muster again after we passed. Could this really be the same cold, grey ocean we had fought so bitterly only a few short months before? Was it possible that in the twinkling of an eye it could revert to type and leave us huddled and shivering behind the canvas storm dodgers? We sat in our shirt sleeves in the wee small hours and trailed our calloused, grubby hands over the side to see the signature they made on the smooth surface. It was almost as if we were taking liberties – like stroking a sleeping tiger.

The cutter lay at anchor, a haven of light and shelter against the black horizon. We went about our business attending outward shipping and boarding inward bounders and as the first grey streaks of dawn appeared in the eastern sky, we might have moved all night without having to disturb the tranquillity of the men asleep on board. It was seldom we needed the cutter under way at times like these but occasionally an incident would occur which demanded some prompt action.

One such incident happened on a clear, calm, summer's morning. Two ships were ordered to sail on the early morning's tide from the LNER Dock. The tide went at about 0400 so the ships were ordered for the first of the gates at about 0130 and 0200. There was nothing unusual in this and we expected to see the first ship, the *Mutla* at about first light, and the second, an Ellerman Wilson ship, would follow shortly afterwards. Both were fast, modern cargo liners and once clear of the Dock entrance would waste no time in the river and clearing for sea.

We, the boathands, had been up and about all night, in around the lighthouse and once or twice up as far as the Red Gas Buoy in the inner reaches. We noticed the fine weather had brought out a handful of spare-time fishermen who seemed to prefer the night's fishing to a warm bed and a good night's sleep. Ah well, it takes all sorts! Some fishermen tended to encroach into the navigable channels and we would drop alongside them for a friendly chat and point out that it would be a good idea to keep clear of the shipping lane. By and large they took this advice in good humour for, as we pointed out, small boats, especially unlit ones, were practically invisible from the bridge of a big ship.

As the night wore on we kept an eye open up the sea reach and eventually we saw the mast head and side lights of a big ship rounding the leading lights and squaring up on its course to the open sea.

With the daylight asserting itself in the eastern sky we jumped into our boat for the umpteenth time, swung the starting handle, cast off fore and aft, and swung her bows round to intercept the Norse Line ship as she cleared the lighthouse at a good eight or nine knots. The grey paint on her hull picked up the first streaks of daylight and the windmilling beam of the South Gare Lighthouse flashed a brilliant shaft across her bridge and funnel. She made a fine sight in the stillness of dawn, her navigation lights shining jewels in a silver crown. Suddenly a white plume enveped her funnel and drifted aft as she blew a long, mournful signal on her whistle. One long – three short. That's for us!

Cracking open the throttle a bit wider we made our best speed towards *Mutla*; at the same time we could make out the dark green hull and white superstructure of the second ship, the Wilson boat, rounding fifth buoy corner about a mile or so astern. 'Apple Sam' on the Wilson boat wasn't wasting any time either, then he blew a long shrill blast, tailing off into a series of hisses and gurgles as some steam whistles do when they haven't been used for some time. There seemed a sense of urgency all of a sudden but we couldn't see any reason for it. Conditions were perfect.

We turned and ran alongside the first ship, laying the boat's bow up against the grey steel plate as the Pilot swung his leg over the gunwale twenty-five feet above us. Swarming down the ladder he arrived breathless in the fore part of the boat.

'Get the cutter underway, son, I've run a boat down in the sea reach,' he gasped. We looked at each other but before we could speak, 'Apple Sam' blew another piercing shriek on his whistle. The familiar signal – one long three short.

'He's probably steamed through the wreckage before he's seen it,' Dan Slater muttered to himself, clearly very worried.

'Shall we take him out first or get the cutter under way?' Peter addressed himself to the Pilot who was straining to see any signs of wreckage. But we were too low and too far away.

'Go to the Wilson boat, son, he may have seen more than me.'

'OK, Mr Slater.'

And we tore towards the second ship at full speed.

Sam had stopped his ship, and the big propeller, half out of the water, turned lazily under the sudden cut away of his high-built poop deck. The ship was flying light and a long white painted ladder hung down her brilliant side. Heads appeared, peering over the high rail and then the familiar figure of Sam Wilkinson in his high laced boots stood full height on the bulwark ladder. Pointing aft he shouted down,

'We nearly hit some silly bugger anchored in mid-channel; we'd better go and see if he's all right.'

Dan Slater breathed a sigh of relief.

'I thought I'd sunk him for sure,' he said.

'The mate on the fo'castle said he was fending off the bow with an oar and we must have been doing fourteen knots at the time! Then he disappeared out of sight under our stern, pushing off the ship's side with his hands. I was sure he was a gonner.'

Little Sam Wilkinson, the most active man I have ever seen on a pilot ladder, came down the ship's side like a zip fastener, arriving with a thump and a big grin alongside his colleague.

'Silly bugger, standing knee-deep in water in his boat, bailing her out, if you please – with the bluddy anchor down – did you hear me blow? Frightened the life out of me.'

'I thought you would steam through wreckage and maybe run down a man in the water,' said Dan, clearly relieved now and becoming more angry. 'Come on, we'd better get the cutter under way and see what we can do for him!'

'It's a good job the lookout man for'ard was awake,' continued Sam, 'I had to go hard over both ways to miss him.'

Both Pilots, shaken and angry, chattered away as we returned to the *B.O.* to get a sleepy skipper out of his bunk and prepare to go back to look for the luckless fisherman.

The skipper appeared, all crumpled and stubble, from his couch, bellowed, 'Heave away,' to nobody in particular, lit a fag and glared at the kettle on the galley stove whereupon it immediately came to the boil right on cue.

Still muttering about the doubtful parentage of certain of the fishing fraternity and how their wives should know better than to let them out of their sight especially at night, he arrived on his bridge and rattled the telegraph just as the anchor clanged down in the hause pipe. Down below, the engineer, equally disgruntled, was slow to react to 'Stand By' and quickly got a second impatient and prolonged jangling, from the telegraph, for his pains.

By now, everybody was up and about and the saloon buzzed with Dan and Sam telling their combined story between great mouthfuls of tea and drags on hand-rolled cigarettes.

'The man was right in the fairway,' Dan was saying, 'Made no attempt to get out of the way. The first I knew was the mate ringing the bridge from for'ard on the fo'castle phone, to say that he THOUGHT we had missed that boat.

'"WHAT BOAT!" I yelled to the "Old Man" who took the message.

'"The boat that we struck as we came round the corner," the mate replied, screaming down the phone.

'"Blimey" says I, and ran to the bridge wing just in time to see this battered boat disappear under the arse end.' More gulps of hot tea – more clouds of tobacco smoke.

The cutter steamed slowly in as the fireman had been caught cleaning his fires. He had not expected to have to get under way at such short notice and had let the pressure drop back in the boilers. As we approached the site of the morning's drama a forlorn scene met our eyes.

There, standing knee-deep in water in a half-sunken, double-ended boat, was big Jim. His engine was flooded and, in fact, under water; his bits and pieces were floating around his legs. One oar, still attached to the thole pin, was smashed off clean, the other was a splintered mess of crushed wood. The tiller was gone and his rudder hung from one pintle. He was standing in a state of semi-shock, going through the motions of bailing out with an old felt hat. Through the fairlead for'ard his anchor rope held him fast in position.

As we steamed up close alongside him he gave us a cheery wave and grinned at us, peering through water-streaked horn-rimmed glasses.

Dan grabbed the loud hailer and gave him a verbal roasting leaving him in no doubt as to where he had gone wrong. He finished up by inviting him on board for some hot tea and dry clothes.

'We'll take your boat in tow and you can beach her to dry out when the tide ebbs,' he finished.

Jim nodded and, splashing about in his stricken boat, he eventually came up with a rusty old knife with which he proceeded to slice through his anchor rope and drift down alongside us. I threw him a boat rope and he made it fast and clambered aboard, water squelching out of his thigh boots which were rolled down just below the knee. As he kicked them off someone thrust a pint pot of hot tea in his hands and said, 'You're a lucky man, Jim, don't you know better than to anchor in the fairway?'

'Aye,' said Jim. 'But my engine broke down at the Red Gas Buoy and I drifted up on the flood until I looked like going behind the leading lights so I dropped the hook so I could work on the engine.'

'Didn't you have a torch in the boat, man?' Sam asked him. 'You should have flashed at Dan as he came round the corner.'

'Well I did, but the torch is no good so I tried to pick the anchor up, but the tide was too strong for me, or the anchor's foul or something. Anyway, I got the oars out, but then he was on top of me, so all I could do was try to row out of the way. The ship steamed over my anchor rope and dragged me into his side. I thought I'd had it. I swear I went through the blades of his propeller. Look at that oar, cut clean through – anybody got a dry fag?'

By this time the poor man was shaking visibly and not just from the cold. We turned the cutter round and, towing the boat gently alongside, Arnold jumped in, wearing his beloved white thigh boots, and, with a large bucket, soon got the level in the boat down to a few inches.

We lent Jim a pair of bunkering pants and an old pair of wellies and he gradually calmed down, his thick lenses steaming up with each successive sip of the scalding brew.

'Aldis lamp!' the skipper yelled as a light winked and blinked from the bridge of the *Mutla* laying all stopped, just outside the Fairway Buoy.

Peter dashed up the brass-bound ladder and grabbed the Aldis from its clip in the wheelhouse.

Clackety. Clack. Clackety. Clack. He replied to the signaller a couple of miles away.

'He wants to know what's happening,' Peter said, after a brief exchange.

'Tell him "Thanks, but everything is OK and no injuries"' said the skipper.

Peter turned and, resting the lamp in the crook of his arm, began sending the message of reassurance to the ship rolling gently in the slight swell as the hull and superstructure glinted in the orange light of the rising sun.

A lamp winked its thanks and a flurry of foam under its stern indicated the vessel's engines set in motion again as she proceeded on passage.

Meanwhile, down on the main deck, our survivor was surveying the damage to his boat which, all things considered, was minimal. He seemed more concerned that his wife shouldn't get to know about his little adventure, especially as she thought he was working an extra night shift.

'I 'phoned her from the foreman's office last night and told her I had to work a double; if she finds out I've been out fishing she'd murder me.'

'It might have been easier all round if she had,' someone muttered.

Climbing back in the boat in his borrowed boots, he let go the tow rope and we prepared to drop the motor boat back into the water. We would have to tow him right round the shallows where the big cutter couldn't go because of her deeper draft. The motor boat splashed into the water and we soon had the engines started and came up alongside him to take his long painter. He turned to us as we made him fast and prepared to move off. With a glance over his shoulder, the full glory of the rising sun reflected in his spectacles, he grinned from ear to ear.

'There was a time, you know, earlier on, when I didn't expect to see that thing again!'

A big dog seal popped its head out of the water twenty yards away, blinked at us and disappeared again.

CHAPTER 19

From time to time during our period of apprenticeship, young men with similar ideas of becoming River Pilots would be sent down to see how we operated.

One or two eventually joined us, replacing the senior lads as they left at twenty to do their sea time. The majority came and were not impressed with chipping hammers, scrubbing brushes, Brasso and the more mundane side to our activities. Possibly they preferred the pin-striped suit of the junior bank clerk or accountant or even maybe direct entry into the Merchant Navy as an officer cadet, and who could blame them. Our working conditions were wet and cold sometimes, dark and dangerous sometimes and downright humbling most times.

One Sunday morning, after several years had elapsed, we were confronted with the skipper holding an obvious directive from the Super. in the office.

'I've a letter here says a young fellah is coming down to the breakwater this morning at nine o'clock. You go and pick him up from the jetty in the boat. He has to spend the day with us seeing how he likes it.'

'Aye, aye, Skip,' Peter and I chorused.

'Where's Arnold this morning? Isn't he here yet?'

'Oh, it's Sunday, Skip,' said Pete. 'He'll be walking from Seaton and we'll have to pick him up from off the Snook sands.'

'Well, wait 'til this fellah arrives before you go up for Arnold, or you might miss him,' said Blood.

'Arnold ain't going to like it,' I muttered to Pete under my breath, 'If he has to stand on those sands for over an hour he's going to be hopping mad when he comes on board.'

Old Blood shrugged his shoulders. He had his letter from above and the senior boathand would just have to wait until we had time to pick him up.

As it was just turned eight we decided to have our breakfasts while keeping our eye on the sands about a mile away and the breakwater and jetty just a couple of cutter's lengths away.

At eight-fifteen a lone figure appeared walking across the sands. Like a lone castaway Arnold plodded a solitary path, his old army haversack over his shoulder and his well worn sea boots rolled down below the knee. It was heavy walking and every now and then he stopped to wave his arms to attract some attention from the cutter. He was nursing a bit of a hang-over from the previous evening in the Marine Hotel and his temper wasn't improved when I answered his frantic waving with a burst on the Aldis lamp to the effect that he would have to wait. After a rude gesture in our direction he resumed his walk along the sands to the spot at the sand edge where we usually picked up and landed the Seaton men. I replaced the Aldis in its clip and returned to my sausage and egg. I only hope it doesn't rain or he'll get wet, I thought. The sky didn't look too promising and for someone on those sands without an oilskin, there was just no shelter. An odd post dotted here and there, relic of some far gone day, was all that broke the smooth sweep of fine sand. It had been said that in bad weather, with gale-force winds and sleet and rain lashing down steel rods, three or four Seaton Pilots would stand huddled in a row under the lee of a single post until the boat arrived to pick them up. Legend had it that in winter time, by the time the relief boat arrived, they had to be prised apart. This may or may not have had a grain of truth in it but I did know first hand that they were never sweet-tempered after a long wait on the sand edge.

Nine o'clock came and went and, with still no sign of our visitor, we asked Blood if we should collect the senior lad.

'No! By jingo,' he replied. 'Tidy up the breakfast pots, clear up the galley and then if he hasn't turned up you can go and get him.'

'Aye, aye, Skip,' and that was that.

We collected the dirty pots and began washing up, keeping an eye open out of the upper half of the galley stable door.

We had just finished tidying away and were busy putting a bucketful of coal in the galley bunker when a lone rider appeared on a motor cycle and stopped at the top of the jetty. This looked like our man. He hauled his bike onto its stand in the lee of the concrete wall, removed his goggles and gauntlets and stuffed them in a haversack and, after a long look in our direction, sauntered down to the end of the jetty.

'This looks like him, Skip,' I yelled down the companionway to the deck below.

'Aye, and not before time. Go and get him then and then run up for Arnold. When you come back we'll get under way.' Blood rolled himself another cigarette and stuck it in his long holder to enjoy with his pot of tea.

Pete and I dried our hands, leapt into the boat and set off towards the jetty.

The youth who had arrived had obviously been briefed on the kind of thing he was coming to. He was wearing a woolly hat, a long black oilskin coat over a navy blue, thick, fisherman's gansey, blue melton trousers were tucked into spotless sea boot stockings and these in turn thrust into shiny black gum boots. Everything he had on was obviously brand new and of the best quality.

'Watch out for that second rung, it's bent, and there's one missing further down.' Peter shouted advice as the visitor turned and made to descend the vertical ladder into our boat. He climbed gingerly down in his stiff new oilskin giving the distinct impression that he was not all that used to wearing heavy sea boots. Clumping into the bottom of the boat, he turned and smiled at us.

'Sorry I'm a bit late,' he said, 'but I had a bit of bother with the bike. It's brand new and I had some trouble finding the gears with these things on.' He indicated his boots, all stiff and unyielding. 'My name's Robin and I have come for the day.'

Peter indicated where he should sit and, smiling again, he half sat and half fell onto the for'ard thoft dragging his shiny, new army pack after him as we backed off the jetty at full chat, our wake lathering the water into a flurry of white foam under our stern. The gear lever was slammed across and the bows swung with the tiller hard over as we headed out from the jetty away from the cutter straight towards the slag training wall.

'Aren't we going to the cutter?' our visitor asked as we headed up the river.

'We have to collect the other lad who's been waiting on the other side for nearly two hours,' said Pete. 'This is a short cut across the slag wall, saves us nearly a mile, see. If we keep the corner of the jetty in line astern with that buoy over there right ahead, we have a narrow gap of about twenty feet with just enough water to sail through.'

'What happens if we don't get through?' Robin asked.

'We hit the bottom with a hell of a crash and you jump over the side to shove us off again,' Peter grinned.

'Why me?' Robin's face fell.

'Because you're the only one with sea boots on,' said Pete and we all grinned our heads off.

Robin fumbled about in his haversack and produced a shiny black sou'wester which he rammed on his head over his woolly hat and sat quietly watching the wake as the engine roared away.

We passed channel-marking beacons set solidly into great square blocks of concrete standing on the slag retaining wall like fortresses. As the river channel is kept at an artificial depth by constant dredging, these walls of slag and rock are dumped along each side of the channel to consolidate the banks and prevent fine sand and silt being washed into the navigable channel. Every so often long wooden piles of greenheart are also driven in to act as marker beacons, each having a distinctive top mark. These form an ideal vantage point for the gawky cormorant bird, standing alone and impassive, often flapping its great scraggy wings as if impatiently drying them in the pale sunlight. In contrast, the elegant and immaculate gulls stand in ranks, head to wind, rising in squadrons as we thundered past close by to become a squabbling, screeching rabble, as they wheeled and dived for the prime perches. I was always surprised at the number of gulls with only one leg. One would think a bird so handicapped, once in the water would swim round in circles, but they don't! They remain calm and unruffled on the surface; mind you, what's going on underneath is anyone's guess!

We motored on towards the edge of the sand where Arnold stood. As we approached, an oar was made ready to use as a quant pole. The tide ebbs fast at this junction of the main channel and Seaton Snook and any delay can result in an embarrassing stranding. Swinging the bows in towards the shallows, we stopped the engine and glided gently onto the rippled, sandy bottom. Arnold was over the bow and into the boat like a flash as we pushed her off again with the oar before she had time to settle. The engine roared into life at a swing of the handle and we backed off and set course for the cutter. The whole set piece lasted less than a minute. We were well practised in this manoeuvre as it was often carried out late at night in total darkness.

Arnold's arrival triggered off a tirade of abuse and obscenities at everything in general which caused our visitor's jaw to sag. The tranquillity of the morning was shattered and so busy was he haranguing us that it was several moments before he even noticed Robin.

'Who's Captain Courageous here?' he finally growled.

Robin, jerked out of his awesome regard of the senior hand, blushed and explained his presence with considerable embarrassment.

Arnold eyed him up and down. 'Expecting bad weather?' and winked at us. An uneasy peace was restored.

The journey back to the cutter was slightly quicker on the ebb and the skipper, with his inevitable pot of tea and cigarette, threw us the rope as we came alongside.

'Pick the boat up, lads, and we'll get underway, we've wasted enough time. Now, son, what's your name?' The last remark was addressed to our newcomer.

'Robin, Sir,' was the reply.

'I don't know where you get all these fancy names from,' muttered Old Blood. 'Do you get sea-sick then?'

'I don't think so.' Robin blushed again, being the centre of attention.

'Well you'll be all right today, it's fine weather, just a little bit of ground swell, go for'ard with the lads and they'll show you how to pick the anchor up.'

We quickly lifted the boat and bowsed the falls hard in with the tackle blocks, then went for'ard to heave the anchor away. Robin stumped after us self-consciously, trying to keep out of the way while still showing interest in what was going on.

We steamed out slowly to our anchor station close to the Fairway Buoy, trailing a long plume of dirty grey smoke behind us. The cutter rolled lazily in a long, almost oily, swell. There was no wind to speak of, it had dropped away to light airs and the swell caused the cutter to move ever so slightly.

Conditions were ideal for a spot of fishing so once the anchor had 'brought up' we hunted in our battered lockers for something resembling a fishing line. They inevitably became entangled and mixed up with leaky condensed milk tins, old torch batteries, spilt packets of tea and rusty ex-army jack knives. It was while this search was going on that one of the pilots on the after deck spotted a large, triangular fin sticking out of the water about two or three cable lengths away. It was clearly moving in our direction in little, hesitant movements. A shark!

In the summer time, when the North Sea warms up a few degrees, sharks are not uncommon, although from the size of its distinctively triangular dorsal fin, this looked a sizeable one. All thoughts of fishing lines went by the board as Arnold dashed aft to find the boat hook. He was itching to have a go at something and that shark filled the bill nicely.

'Put the boat down quick, and don't make a row.'

All the Pilots came out on deck to see the fun. Robin, looking a bit pale, asked what we proposed doing.

'We're going shark fishing,' I retorted. 'Fancy coming, it's just a bit of fun?'

'Wh'what kind of fun – aren't sharks dangerous?'

'Hell no! These are basking sharks, they won't eat you. We're just going to give it a bit of a dig in the ribs with the boat hook, that's if we can get close enough. Come on, you'll be all right.'

Arnold arrived with the ten foot boat hook. It had a wicked looking six inch galvanised spike on the end of it and, brandishing it, he jumped into the fore part of the boat.

'Let's go! Let's go!' he giggled excitedly.

I pushed off from the rusty side of the cutter and motored gently round the stem to the direction indicated by a group of Pilots standing right aft. The new lad sat nervously alongside me in the stern sheets as Arnold stood right for'ard grasping the painter in one hand, the boat hook, like a harpoon, in the other, his feet astride, one on each side bench, as he leaned slightly backwards against the forward impetus of the boat. We approached the fin with all the stealth we could manage.

'How big is it, d'you reckon?' I whispered to the intrepid harpooner.

'Can't tell yet, can only see its fin, but it's swimming all the time. Not so fast, you'll scare it.'

I popped the gear lever in neutral with the engine just ticking over, and, pushing the tiller hard over, tried to lay the bow of the boat as close to that fin as possible.

Out of the corner of my eye I caught sight of Robin gripping the gunnel with whitened knuckles. His face had a distinctive anxious look and a paler, greenish tinge.

There is something unaccountably awe-inspiring when you are in the presence of a large creature of the deep. I think it is the air of mystery which surrounds them, never quite knowing what to expect.

The harpooner steadied himself, aimed, and then lunged down over the bow with the heavy pole.

The effect was instantaneous and electrifying.

A great tail fin rose in the air and slashed across the stern sheets of the boat, knocking the tiller clean out of my hand and almost slapping Robin across the face, soaking us both to the skin and half filling the stern of the boat with a flurry of foam. Robin yelped with fright. Arnold collapsed in a heap on a coil of rope for'ard as the whole boat was flung sideways two or three feet by the enormous beast under it. I gasped in amazement at the size of that shiny black fin and the sudden sting of cold water. The creature must have been well over twenty feet long. We caught a brief glimpse of a great, broad black head and then it was gone in a flurry of agitated water.

'Nearly got it,' said Arnold in utter bravado.

'Tell you what,' says I, 'If that thing had jumped into the boat, I'd have been over the side pronto.'

Robin, wet through and shaking visibly, had turned even whiter. A great cheer rose from the bunch on the cutter's after deck as Arnold retrieved his weapon and we motored back alongside. We made the boat fast with a bow line and stern line as Robin, making himself as much part of the scene as he could, pumped vigorously on the rotary bilge pump handle. It took several minutes to pump the boat dry and normally we would have just lifted it clear of the water and removed the plug but we thought it would take our guest's mind off things if we let him do something useful rather than just stand around like a spare part.

By now it was time to eat. The usual scramble developed for frying pans and other utensils. For this reason I often brought just a jar of home-made soup or a cottage pie, something that just required heating up and was not ruined if left to its own devices if we had to attend to our shipping at meal times. If we looked like being some time we just took a pint pot of soup and a hunk of bread in the boat with us.

Robin offered us his salmon and cucumber sandwiches, saying he wasn't very hungry. Salmon and cucumber! This was birthday party stuff! We dived into his sandwich box with all the finesse of gannets being fed with sprats. Mind you, I always found cucumber (together with lemon curd, but not at the same time) a bit indigestible, so I carefully picked out the cucumber rings and gave them to Pete or Arnold.

The Pilots of the watch were brewing up a great big black pan of scouse (or thick soup). Someone had brought a ham shank, others potatoes, carrots, leeks, split peas and lentils and the galley was full of shirt-sleeved Pilots peeling, scraping and chopping. We couldn't get near the place to do any cooking, we would just have to wait until the preparations were complete and the mixture simmering away. The smell coming from the galley was almost unbearable! Onions, garlic, a dash of curry. We would nip along with a couple of pint pots and, when the coast was clear, quickly ladle a couple of helpings, top it up with boiling water and beat a hasty retreat.

What with salmon sandwiches and No.6 Watch's scouse, we were doing well today. Not so Robin; he was starting to feel decidedly queasy. All the symptoms were there: palish green pallor, clammy sweaty skin, lack of interest in anything around him, constant need of fresh air, lack of interest in highly spiced ham shank scouse. Poor Robin, he was putting on a brave face but King Neptune's secret weapon was taking effect. Would he last the day?

The skipper came on deck carrying his empty soup plate, took one look at the unfortunate lad who was leaning over the side, pretending to study the surface of the sea, and shouted for one of us.

'If that lad's feeling ill and wants to go ashore, take him with you next time you go that way.'

'Aye, aye, Skip,' said Pete, 'But there's no more sailings 'til relief time.'

'Well, there's no need to make a special journey, he hasn't started feeding the fishes yet, has he?'

Old Blood's gentle upbringing occasionally showed through in his ability to avoid the unpleasantnesses of life in the raw, by an avoidance of plain speaking.

'No, he hasn't spewed up yet,' said Pete, calling a spade a spade.

'Ah well, he might be all right, try and find him something to do. Take him in the boat with you every time.'

'Aye, OK,' Pete said, and Blood disappeared below to his cabin to conduct his afternoon's deck head survey in peace.

We set about washing up the empty soup dishes and pans. I forgot one ingredient that they must have used, maybe secretly to teach us a lesson, but certainly effectively when it comes to washing up – glue.

When Tommy Dee, the senior Pilot, said his soup would stick to your ribs, he wasn't kidding. It stuck to everything, and trying to clean the big iron pan was murder. Holystone, bathbrick, Vim, the lot, it was tenacious, we even tried sharp sand and a scrubbing brush. Eventually, we won and restored some kind of order to the galley while Arnold, being senior, snored his head off in the fo'castle.

We had just finished in the galley when a couple of blokes in a small double-ended boat appeared alongside. They said they'd found a 'floater' about a mile away, in Coatham Bight. They'd managed to pass a line round 'it' but for some reason or another wanted to pass on the responsibility to us to tow 'it' in.

Now a 'floater' is the term used for a body found floating in the sea. They have usually been in the water for some time and the natural processes of decomposition have caused gases to form internally thus giving them a degree of buoyancy causing them to rise to the surface. They weren't pleasant things to have to deal with but the fact that they had managed to pass a line round 'it' made it more acceptable. You were paid by the local authorities for landing a body, ten shillings on the Yorkshire side and, for some unaccountable reason, fifteen shillings on the Durham side. So, for obvious reasons, we always tried to land them on the more profitable side. We agreed to look after it and they gave us a rough position with a bearing and distance and said they had tied it to a fishing dan or marker buoy so it would not float away with the tide. We called the signal station at the lighthouse with the news and requested a suitable means of transport. Everything arranged, we were about to set off when I suddenly remembered Robin who had sneaked away into a quiet corner to feel green all on his own!

'Come with us in the boat,' I said, 'and if you don't want to stay the rest of the day, we'll land you back at your bike. Maybe you'll feel better in the boat.'

Robin accepted the invitation with a wan smile. We tried to jolly him along as we set off, not telling him where we were going as we thought it might upset him.

Motoring in towards the beach we soon sighted a raggy bit of bunting fluttering from a long bamboo pole that marked one end of a fleet of crab pots. The men in the double-ender had followed us to make sure of the rendezvous and indicated with shouts and gesticulations when we reached the spot. Sure enough, when we detached the old piece of tarry hemp from the cork slabs around the marker pole, we had an old lady fully dressed including a straw bonnet. She was wearing gloves and around her wrist was a handkerchief tied securely to her handbag. She was bobbing up and down gently in the wavelets, floating characteristically face down. Robin asked us what it was as indeed 'it' looked for all the world like a bundle of old clothes. When we told him he almost passed out, and immediately went green again.

'What are you going to do with her?' he managed to gasp.

'We're going to tow her ashore. We'll get ten bob for her, maybe fifteen if we can make Seaton Snook with her. Take a turn through that ring and give her plenty of slack,' I said, indicating a large metal ring in the stern of the boat. 'It looks to me as though she's jumped in, the way her handbag's tied round her wrist. Poor old thing, maybe even just walked in. Either way she seems to have been prepared for it. Shame.'

We set the boat off gently. As the rope came tight we throttled back. The body began turning round and round on the end of the tow rope.

'Take it easy, Pete, we've a long way to go yet, and we won't be able to tow her at this speed.' We throttled back to a tick over.

'How's she doing now, Robin?' I asked. 'Can you see?' hoping to involve the lad.

With a great effort of self discipline, Robin gulped and belched and looked aft to where our body was gently windmilling round on the end of the rope just clear of the propeller wash.

'She – it's – she's still going round and round,' he gasped, and quickly looked away.

'Oh, hell,' said Pete. 'Take it out of gear, and we'll go as slow as we can. If she's been in the water a long time she might start coming to pieces on us. Then we'll have to start searching round for the bits!'

This last remark finished Robin off completely and he was as sick as a dog. I quickly grabbed him and indicated as delicately as possible that he should deposit all over the side.

'Don't do it in the boat, old son, we'll have to clean it up.' Besides I could still taste that cucumber and my stomach wasn't all that cast iron after Pete's tactless observation.

Even Pete realised what he'd said and probably wished he hadn't. He pulled a face and was just going to say something when the Aldis lamp at the lighthouse keeper's watchroom winked away at us.

'What's he say?' said Pete, as I waved my arms in acknowledgement to each word spelt out in flashes to our boat.

I shushed him angrily as his remarks broke my concentration on that point of winking light. Waved again and turned to him.

'Says transport has arrived at the Gare and do we need any assistance.'

'Damn, that means we'll have to land her at this side.'

'Well it'll take long enough with the way the tide is and towing at this speed without trying for t'other side.'

We cut as many corners as we dared and eventually, still escorted by our friends in the double-ender, we arrived at our own Government Jetty to see a black ambulance-type vehicle parked at the top of the walk-way and several uniformed figures peering over the wall at us. Running close up to the shore the bows ground to a halt on the pebbles; we unhitched the tow line and passed it for'ard, coiling it in readiness to throw to the big policeman who had appeared on the shingles.

'Now, my lads, what have you got here for us?' he rumbled in an official voice. Taking care not to get his highly polished boots too near the lapping water, he gingerly accepted the length of rope offered him by Robin who had recovered somewhat and only looked white now.

'I think she's all there!' shouted Peter from the stern of the boat.

'Right-o, me lads, we'll take the remains from here,' and he towed the pathetic little bundle into the shallows as far as he dared without getting his boots wet.

Two others appeared carrying a black tin coffin and marched straight down the shingle, setting the coffin down. The lid was lifted off and, after a bit of fiddling about, its sides collapsed flat.

It was obvious that these last two chaps were used to such situations and ready for this one. Wearing gum boots and rubber gloves they waded into the shallows and, after untying the tow line, they rolled the sopping bundle over the flattened coffin side and once in, lifted the collapsed side, secured it and replaced the lid. One of them, the taller of the two, turned to Robin who had by now climbed over the bow and splashed thankfully ashore.

'You look a bit pale, son, would you like a wiff of oxygen? – we've got some in the wagon.'

Robin declined and, with a wave over his shoulder, trudged up the beach behind the sad little procession.

That's the last we ever saw of Robin. Someone, much later, heard he'd joined the Penny Bank. I don't think he'd ever forget his day spent with us. I often wondered, later on, whether he imagined all our working days were spent like that Sunday.

As we backed off the shingles to return to the cutter, our double-ender escort shot out from behind the jetty where they'd been lying.

'Wonder what he's hanging about for.'

'Maybe he wants his tow rope back,' and I coiled up the length of rope we'd been using.

As the boat came close to I yelled across to them, 'D'you want your line back?' and waved the coil at them.

'No, you can keep it.' One of them wrinkled his nose in mock distaste. 'We just had a twinge of conscience when we lumbered you lads with the job. Thought we'd stay handy in case you needed us – so long – we're off back to our pots now.'

And with a wave they peeled off and were gone.

'You know,' Pete turned to me, 'I don't think I'd fancy a crab out of their pots for quite a while.'

CHAPTER 20

Every year, usually towards the beginning of July, the time came for the steam cutter's annual boiler cleaning, dry docking and general overhaul. The winches were getting tired and leaked steam and oily water, the grass grew long and green on her rusty bottom and everything had a well-used look about it. It was on these occasions that the Pilotage Service became shore based as the Pilots moved into the brick-built building at the top of the jetty, known to everyone as the shore cabin.

This consisted of two rooms, one of which was the watch room with a table and wicker arm chairs. Around the walls, in two tiers, were iron bunks. A large stove against one wall provided the only heating, and communications was a telephone to the switch-board at the lighthouse. If the lighthouse watch-keeper chose to ignore us or, more likely, was somewhere else when we rang through, we had no telephone contact. The other room was the galley with a trawler-type solid fuel stove, an enormous pot sink and a single cold water brass tap. A toilet and fuel store completed the amenities. The floor covering throughout was battleship-type linoleum, and the windows were clear glass and curtainless. A flight of narrow twisted wooden stairs led directly out of the watch room up to a look-out tower with sliding windows. It was in this shore cabin that the watch of Pilots spent their time waiting for shipping to arrive. We, together with the skipper and engineer, lived aboard the diesel cutter *W.R. Lister,* moored at the end of the jetty.

Built at Anstruther in Fife, hence the name 'Scottish Fifer', the *Lister*, as everyone called it, was stoutly built of larch on oak with extra deck beams to strengthen her for going alongside ships. Instead of a fish-hold it had quite a roomy saloon with side settees and six bunks. Right for'ard in the extreme fore part was the galley, cramped and with very little headroom. Down below decks aft was the lads' accommodation shared with the engineer. The engine room access door, which was kept constantly wide open, led directly out of the after cabin. We got quite used to sleeping with a seventy-five horse power Paxman Ricardo Marine diesel engine thundering away about ten feet from our bunks. The wheelhouse, partially divided into two by an athwart-ships partition, sat on top of all this and a short companion way or staircase led down for'ard and aft from the respectively partitioned parts of the wheelhouse. The for'ard wheelhouse contained wheel, compass and engine controls and the after wheelhouse had bench seating and a small coal bunker-cum-writing desk for the pot-bellied stove in our cabin down aft. The whole craft was snug, smelly and very seaworthy but the thing we liked about it most was that we didn't have to turn out for every ship we attended. The craft was designed, built and heavily fendered to go actually alongside ships to embark and disembark Pilots. And as such the skippers, not us, did all the work, and we merely made ourselves useful generally, running the cutter, tying up and attending ladders or keeping out of the way.

At the time of change-over from one way of working to another, chaos reigned. The shore cabin had to be cleaned from top to bottom. Bedding had to be cleaned and aired and the place made reasonably habitable after the near-neglect of months. Likewise the *Lister*, although thoroughly seaworthy in most respects, had often just undergone a complete mechanical overhaul and, as a result, tended to be dependable but dirty. Maintenance engineers are a notoriously greasy lot and they are inclined to wreak havoc with white enamel or eau-de-nil gloss, especially in the saloon and galley. The aft accommodation, where we had to live, had usually been used as an annexe to the engine room. Except where the engine room gleamed with polished brass and copper and freshly painted, lovingly burnished machinery, the aft cabin was littered with greasy paraphernalia. Oily waste, five gallon drums, coils of rope, tools, anything that would detract from the twinkling, winking, Aladdin's cave of an engine room was hurled into the after cabin to be cleared up at some future date. A strong smell of diesel fuel and oily bilge water pervaded the place.

We had to set to and make the place fit to live in and this involved many hours of scrubbing and washing paintwork. This general upheaval of our normal working system upset our working partnerships. Several lads were required working days on the steam cutter up in the dock and those left were usually split up into watches of just two. This summer arrangement meant we worked with several different lads as we took our summer holidays during this time and worked twenty-four hours on and twenty-four off for a period of eight or ten weeks. By the time we had been on twenty-four hour watches for eight weeks we were ready for our annual fourteen days leave. My working companion this particular time was Geoff Mulligan, an irrepressible comedian, who often had differences of opinion with the skipper owing to his constant sky-larking.

Anyway, in no time at all, the skipper had us boiling buckets of water on the galley stove, necessary for washing all the interior enamel work. We worked sweating, stripped to the waist, with great blocks of soap and wads of mutton cloth, washing out between the deck beams and the deck head stiffeners, the soapy water running up our arms and down our torsos in long rivulets. Bucketful after bucketful became filthy greasy and was changed for fresh, and gradually, after several hours, the forward saloon and particularly the galley, showed the result of our labours. This was Captain Jack's domain where he lived and relaxed and it had to be CLEAN. By the time we had spent a whole day on it, it shone, and we were filthy and exhausted. Jack Emmerson grunted his approval, finally, at about afternoon 'smoke-o' and told us to clear up for the day. One final bucketful on the stove was for us to get a bath in down aft. After a meal of beans and bacon and a pint pot of black tea we felt much better and even managed a bit of good natured banter with 'Old Parky' the engineer.

Parky, a wizened little man, was long past retirement age but came down each summer to help us out and probably to keep out of his wife's way. He had sailed all during the war in tankers and he sat down aft for hour after hour smoking a battered old briar and spitting expertly at the stove which he kept at a constant cherry red glow. He said he'd spent so much time in engine rooms in the tropics that anything below 90°F and he started to suffer with the cold. Be that as it may, he always slept with all his clothes on and a woolly hat pulled right down over his ears. Ignoring all our protests, he rose, from time to time, for a cough and a spit and, rattling and riddling the old stove, proceeded to stoke it up until the atmosphere down aft was almost unbearable. Woe betide anyone who suggested opening the skylight ventilator. It stayed firmly battened down 'in case we shipped a sea, son'. We suffered and tried to get to sleep in the hot, oily atmosphere.

Daylight hours being long, we usually set watches about 9 p.m. and, depending on how busy we were, would call the second man in the early hours, say one or two o'clock.

It was the job of the man on watch to tend the moorings and see the boat didn't get hung up on the falling tide, also to keep the galley fire in good shape in case kettles were required during the night and, lastly, to keep a lookout for ships sailing so he could warn the skipper. The Pilots at the top of the jetty in the cabin kept a watch for ships arriving. Skipper and engineer usually turned in about midnight and hoped for a quiet night. Sometimes they were lucky and got four or five hours, more often than not, they weren't.

We were usually so tired by one or two that we just fell into the bunk the other lad had staggered out of and were asleep in no time. We left him to creep around as best he could to avoid waking the other sleepers. Invariably the first stop was the galley, tiptoeing past the snoring skipper, to make a quiet pot of tea without disturbing him. Once made, I would often take my drink up onto the jetty and drink it, leaning against the rail. During the summer, there were often all-night fishermen tending their rods and a pleasant half hour could be spent in small talk.

I gave Geoff a shake about 1.30 and he woke with a start, banging his head on the heavy wooden deck beam knee a foot above his head. As he feigned unconsciousness I poked him again.

'What's up? What time is it?' he mumbled, his brown face crumpled with sleep.

'It's getting on for two and everything's quiet,' I hissed.

'Put the light on, then, so I can find me boots.'

'Don't wake Parky,' I said, glancing at the huddled heap in the opposite bunk. The Chief was snoring loudly and his false teeth rattled with an odd, clattering vibrato.

'Blimey, you won't wake 'im,' grumbled Geoff, 'Once he's away, you could peg 'im out on a clothes line and 'e wouldn't wake up.'

Geoff crawled sleepily out of the warm bunk as I surreptitiously cracked open the skylight for a much-needed breath of air in the place, took off my boots, slackened my belt and slid into the snug nest of ex-army blankets that Geoff had just left.

'Watch the moorings, Geoff, the tide's falling fast,' said I as he climbed the five or six steps into the after wheelhouse clutching his sandwich box.

'Aye, OK, I'll go and have a read in the galley 'til daylight, then I'll do the decks,' and he disappeared down through the for'ard saloon into the galley and once inside, closed the flimsy door behind him.

It had been a hard, long day but the *Lister* was a whole lot cleaner. I closed my eyes, the stove crackled, a cool breeze fanned my face and I was in the land of nod.

The galley was placed in the very bows of the cutter and, as such, almost came to the point of the bow. The only thing for'ard of it was the anchor cable locker and this was arranged so that its lid formed a seat. It was possible to sit with a degree of comfort resting your back against the inside sweep of the bow while your legs and feet braced you against the curve on the opposite side. This was a favourite place, especially in lousy weather, as it was very snug if not claustrophobic. The skipper didn't like you spending too much time for'ard as, obviously, a good look-out couldn't be kept, but when he was asleep, we often crept for'ard, opened the little deck head hatch to listen for tell-tale sounds such as ropes creaking or a ship blowing, put the light out and watched the glow of the stove. This stove was small and placed up against the wooden partitioned bulk head backing onto the skipper's bunk. A zinc draining board and a small metal sink with a manual fresh water pump, made up the rest of the facilities. If you were taller than about 5ft.7ins. you had to stoop to do anything in the galley, and if you were unfortunate to be caught cooking or washing up when the cutter went to sea, it was almost impossible to keep your feet in anything but the mildest weather.

Anyway, it being a fine night, Geoff settled down with his tea and a sandwich, to while away a couple of hours before dawn. The stove was drawing well as an off shore breeze had sprung up and the cutter lay quietly, just giving the occasional jarp against her fenders. The slap of the halyards against the mast, the sigh of the wind, the tickling crackle of the stove – he nodded off!

CRACK! He woke with a start. The cutter lay at an unfamiliar angle. It was broad daylight. The stove was cold and black.

'Christ, we're hung up,' Geoff heard himself shout.

'What the hell's going on!' Captain Jack was a stirring, angry bear awakened rudely from hibernation.

'It's OK, Skip, it's OK,' Geoff yelled as he shot past him on his way up on deck. He and I collided as we arrived in the wheelhouse door together, and scrambled on deck. Luckily the ropes hadn't jammed on the bits so we were able, with care, to release the tension on them without anything carrying away. The old cutter jerked and sloshed and wallowed back into a state of repose again without any apparent damage. Not so Captain Jack; he was on the warpath. He'd fallen off his bunk and landed on his elbow, picked himself up expecting to be scalded to death by the kettle which had bounced off the stove top but instead found himself with a lap full of cold water.

'The bluddy stove's out!' He roared so loud, I'll bet his wife in Seaton heard him.

Geoff, partly regaining his composure, dashed into the engine room and, emerging seconds later, almost fell over old Parky, who was surfacing by degrees. Clutching a fist full of soaking cotton waste, Geoff hurried down for'ard past the ranting skipper.

'What the hell's going on?' Old Parky said, fumbling for his pipe and matches.

'I think Geoff's let the stove out,' said I. 'He's taken some paraffin waste to relight it before the skipper throws a wobbler.'

'He'll have a job,' Parky grinned, applying a match to the dregs in his briar. 'There's no paraffin in there, only a can of petrol.'

'Oh no! – Geoff don't try to -' I yelled – too late. SWOOOSCH. The noise sounded like the *Flying Scotsman* starting off in Darlington station.

A great black cloud shot out of the galley stove pipe on the fore deck, followed by a yard of orange flame. It developed a mushroom top and then fell, completely enveloping everything.

But below decks, every single vestige of soot, ash, clinker, coal, cinder, that had been neatly contained in the cast iron confines of the stove, was now liberated into the atmosphere of a much larger space, namely the galley and saloon and in amongst it all, coughing and choking, were Captain Jack and the luckless Geoff, now minus eyebrows and a goodly part of his hair.

Particles of ash and soot remained suspended in the air for several seconds and when they eventually began to settle and it became possible to see again, the dimly grotesque figures of Skipper and lad could just be made out, covered from head to foot. Jack was so mad he was speechless. Geoff, lucky not to have been badly burned, was shaking with fright and shock. Both were completely unrecognisable, almost invisible, as they blended perfectly with their background. The whites of their eyes and the shadows they cast in the pale dawn light were the only indication of their presence.

Parky was the first to see the humour in the situation and held his aching sides as he rolled around in uncontrollable mirth.

Geoff, with great presence of mind, sensed he was in some physical danger, dodged past the Skipper in the gloom and headed for the iron ladder on the jetty. He went up it so fast, his feet scarcely touched the rungs, with the bellowing of Jack Emmerson ringing in his ears.

'Mulligan! So help me, I'll swing for you yet!'

CHAPTER 21

If Captain Jack was a man's man, Geoff Phelps was more a gentleman's gentleman. Geoff's main involvement on the Pilotage Service was with the *Lister*. He skippered it and engineered it, he got it very dirty and cleaned it, only to get it dirty again. Although not a professional seaman in the same sense that the other two skippers were, nevertheless, he had been knocking about boats most of his life, and was nothing if not versatile in all matters to do with the diesel cutter. In the summertime he arrived to do his stint, as it were, as part of the summertime set up. A small, almost slight, wiry, little man, with swarthy, aquiline features and lank, black hair, well Brylcreemed, he would arrive for his watch carrying an enormous canvas rucksack full of tools and other items vital to the cutter's maintenance. It was said that after a long bad-weather watch during which Geoff had been on his feet most of two days, he was stopped by an over zealous Customs' man in Dock Street, who demanded to see the contents of such a heavy bag. Geoff smiled wearily and obligingly slung it off his aching back right on the man's foot! He limped for nearly a week and I can believe it.

As a skipper, Geoff was very cautious, some would say over cautious. He lacked the swashbuckling panache of Jack Emmerson but had more idea than the older Captain Blood, who was very uncomfortable in the *Lister* and let everybody know it. He suffered at the hands of the more roguish faction amongst the Pilots who took advantage of him at every opportunity. However, he always conducted himself with a great deal of dignity and charm and, I think on balance, the micky-takers were the losers in the end. Whenever they arrived on board his cutter, the Pilots were always greeted with cheerful courtesy and, as often as not, offered a boiled sweet or a mug of cocoa. He was a conscientious servant of the Cutter Company for well over thirty years and, when he tragically died just before retirement, he left a gap which has never properly been filled.

However, during our boathand days, we seemed to get on fine with Geoff. After Captain Jack or the irascible Blood, he was a bit of a soft touch. We knew how far we could go with Geoff and it was a hell of a lot further than with the other two.

Geoff was at his point of greatest discomfort when confronted with a bit of bare-faced contrabanding. He sometimes found himself in the position of actively participating, albeit unwillingly, by virtue of the fact that he was lying his craft alongside foreign-going ships. This, coupled with the fact that some of the older and more unscrupulous types of Pilot on the Service adopted a very cavalier approach to the procurement and consumption of a duty-free bottle, added to his discomfort.

The antics of blackmail, bribery and sometimes plain bullying that the said unscrupulous ones indulged in in order to obtain 'a drop in the bottle' or 'a plug of Nosegay' or 'Scotchcake' had to be witnessed to be believed.

'Show him *me*, Dick,' or 'I'm the Burgomaster, Captain,' were phrases bandied around, often drawing blank expressions from the Captains of very small coasters who happened to arrive when a thirst was running high or the baccy pouch was getting low.

Two irrepressible old reprobates, working together, were Dick Franks and Jack Nixon. Both were well into their sixties and had an uncanny nose for a bottle of Dutch gin or Four Bells' rum.

So it happened one summer's evening, the two were sitting in the shore cabin with the prospects of twelve hours or so ahead of them and nothing in the way of distraction. No chance of a pint in the Warrenby Club as it was too far for them to walk and they were too mean to pay for a taxi. A little Dutchman, the *Jan*, loaded with scrap from the continent, popped up unexpectedly. Neither of them was on turn but here was a chance to wet their whistle and maybe even a cigar or two.

'We'll come off with you, son.' Jack addressed himself to the junior Pilot whose turn it was.

'Mebbe when you tell him *I'm* on the cutter, he'll invite us aboard. Anyway just tell him I'm the Burgomaster, he's sure to want to meet us.'

The junior, not having had a Licence too long, wasn't quite sure what to think at this statement from a Pilot thirty-five years his senior.

'Aye, OK, Mr Nixon. I'll go down and get Geoff ready on the cutter,' thinking to himself things must be pretty desperate to get the two least mobile men on the Service to stir themselves clambering about on ladders and cutters. They were two very big men, in bulk rather than stature; in fact they and two other slightly lesser mortals had earned themselves the irreverent nickname 'the heavy arsed watch'. They took some stirring but the situation warranted it.

As the young man hurried down the wooden jetty the two conspirators togged up in their long gaberdine macs and with greasy, old uniform caps rammed over their wispy, unruly hair. They waddled down to where the cutter lay quietly in the evening sun. Jack, the most incongruous of the two, with his crumpled uniform pants at half mast, his stained blue mac, frayed and tattered round the bottom edge, as if drawing attention to his maroon socks and brown boots, arrived last down the ladder.

'When you go on board, Jimmy boy, tell the Captain that the Mayor is on the cutter and would like to welcome 'im personally like to our port, then I'll nip over the rail and give 'im a real civic reception, so to speak.'

'Righto, Mr Nixon.'

'Maybe you should 'ave 'ad your Chain of Office, Jack,' chipped in Dick, who loved to get a dig in. 'Then maybe you could've given 'im the Freedom of the City, so to speak.'

'Don't be silly, Dick,' retorted the indignant Jack, his pink jowls stubbly with a day and a half's growth of white whiskers. 'You 'ave to give it back at the end of your turn.'

Dick's eye gleamed. 'Aye, but knowin' you I'da thought you'd 'ave figured out a way of keeping it!'

'Now then, Dick, let's be serious. I'm just givin' young Jimmy 'ere 'is instructions. Are you comin' aboard wi'me?'

'No, I'll let you do all the receivin', you've more practice than me,' he said, leaning heavily on the cutter's guard rail, puffing and panting from his exertions.

As the *Lister*'s diesel spun over on air and then clattered into life, Geoff dropped the for'ard wheelhouse window and, with a large boiled sweet tucked into his cheek, shouted, 'Righto, lads, let go.' They backed away from the jetty and swung the bows round towards the river entrance. As the cutter's bows lifted slightly on the long, lazy swell, Jack, out on the cutter's deck, walked round to the open window at the front of the wheelhouse, leaned inside and presented his whiskery face to Geoff at the wheel.

'Now then, when we get alongside, Geoffrey, don't be sheering off too quickly, 'cos Jimmy here is goin' to present my compliments to the Captain, and I'll probably be invited on board – matter o' courtesy, so to speak. Mr Franks 'ere is goin' to be 'andy on the rail and you take your instructions from 'im while I'm on board. Is that clear, Captain, eh?' Jack leered at the unfortunate Geoff who hated these situations.

'Couldn't be plainer, Sir. Aye, aye, Sir.' Geoff sucked furiously on his mint-o, while trying to adopt an air of nonchalance. He began singing in a toneless off-key voice.

The little ship, totally unaware of the reception awaiting it, turned round the Fairway Buoy and headed up the main channel towards us. At one halyard fluttered the plain yellow 'Q' flag and on the opposite side the striped yellow and blue 'G' indicating he required a Pilot. In the wheelhouse the Captain spotted the *Lister* heading towards him and wound away at the large brass throttle screw which protruded out of the wheelhouse deck to waist height, the engines agitated, plonk-a-plonk-a-plonk, subsided a little, and the little ship slowed down as the cutter described a wide arc round to port and came up alongside the rail on the starboard side. It was just a step from the cutter's deck across the tiny side decking onto the heavily tarpaulined hatch. Young Jimmy took it in his stride but the larger, heavier Jack was right behind him, waving and grinning at the wheelhouse, brandishing a couple of old newspapers he'd managed to dig up. The Dutch Captain was getting the full treatment and old Jack positively oozed charm and bonhomie as he skipped into the small, overcrowded wheelhouse. After much waving of arms, hand shaking and back slapping, he and the Captain descended the short ladder and disappeared into the tiny saloon with its neat net-curtained windows and profusion of trailing plants. Geoff, on the cutter, leaned out of the wheelhouse window and fidgeted. Dick, still draped over the safety rail, was alert for any sign from the *Jan*. After several minutes, the grizzled head of Jack appeared, an enormous cigar clenched in his dentures. He removed it to wave excitedly to his colleague.

'See if you can find a bottle, Dick, he's got best part of half a demijohn of Jonge Geneva down 'ere. Says he'll give us a drop if we 'ave sommat to put it in!'

''ere that, Captain,' Dick turned to Geoff at the wheel. 'Find us an empty bottle and we'll 'ave a drop o' gin.'

'I'm sorry, Mr Franks,' Geoff flushed a little, 'I'm afraid the only bottles we have here are milk bottles and they aren't very clean.'

'Well, get the lads to wash the buggers out and fetch us a couple 'ere. Look sharp, man, we 'aven't got all day.'

Two old milk bottles were rescued from the gash bucket, rinsed under the tap and handed up through the hatch to the old Pilot on the rail.

''ere, Jack, 'ere,' shouted old Dick, waving the bottles at him. 'Get one for me, Jack, while you're at it,' and, chuckling with anticipation, he handed the two empties to a boy who had clattered along the decks in battered old clogs.

Several more minutes elapsed and by now the two craft side by side were very nearly abreast of the lighthouse. Geoff began fidgeting even more and the young Pilot came out of the wheelhouse and went down into the saloon to see what was afoot. One of the net curtains was raised momentarily and the tousle-haired figure of Jack Nixon could be seen, glass in hand, flourishing a cigar as though conducting an orchestra. He waved to the cutter, the curtain dropped. Obviously he had struck gold and was making the most of it while it lasted.

At last they emerged, the Dutchman and Jack, both a little glassy-eyed, smoking big cigars. The pockets of Jack's raggy, old mac bulged and sagged. The top of one milk bottle protruded, corked with a tightly rolled wad of old navigation chart. Jack, his hat rammed back on, waddled across the deck, his coat barely held close by a single button, and made to climb up onto the old lorry tyres used by the cutter for fendering. Two young Dutchmen, highly amused at the whole affair, helped him up as best they could. His teeth clenched on his cigar as he muttered and chuntered, he trod on the hem of his coat and a bottle slipped out of the pocket, smashing on the deck.

A look of dismay on his face as he turned round to see the gin swilling in the scuppers.

'Oh, gee, Dick,' he wailed, 'There goes your bottle!'

He heaved himself up onto the *Lister*'s rail, and, with a wave, indicated to Geoff to sheer off.

Crestfallen, but not totally demoralised by the sudden and tragic loss, the two conspirators toddled down aft to discuss the day's events. Jack produced a badly bent cigar for his friend and they both sat in the after wheelhouse looking out over the cutter's stern, taking stock of the situation.

If they went ashore with their gin they would have to share it out with five or six others, plus the fact that they would be liable to prosecution for smuggling and so, they reasoned, it was best to drink it on the cutter. Geoff wouldn't want much, maybe just one little taste. Parky was an unknown quantity, but it was a risk they had to take, so the two decided to sup it there and then.

'D'you fancy a little drink then, Captain?' Old Jack addressed his remarks to Geoff in the wheelhouse, who was greatly relieved at being on his way again.

'No thanks, Mr Nixon, I'd rather not if you don't mind.'

Jack smiled again and withdrew the remaining grubby, Northern Dairies bottle from his jacket pocket. The makeshift stopper had held and it was very nearly full.

'Maybe the Chief would like one,' volunteered Geoff, knowing full well that the wizened old Parky, given half a chance, had an awesome capacity for gin, despite his size.

'I was just about to ask 'im,' Jack said huffily, and then he bawled down the for'ard companion way, 'Fetch us some pots up, son.'

'Let's get it doled out, Jack, before you drop that bugger as well.' Dick's concern was a real one as his companion was becoming increasingly unsteady due to his sampling session on board the *Jan*.

'*I'll* do the honours, Dick, *if* you don't mind.' Jack waved the bottle round as if asserting his command of the situation.

'Would you like a little drink, my old friend?' He addressed his remarks to Parky, who had bobbed up out of the after cabin and was standing head and shoulders in the hatchway. As if to confirm his acceptance of the invitation, he plonked his chipped old pint pot on the top step of the ladder, removed the pipe from his mouth and grinned a big schoolboy grin, nodding vigorously.

'Very civil of you to ask, Pilot,' he gushed.

Jack eyed him with some suspicion and then smiled disarmingly.

'We're just going to have a little one apiece and then give the lads in the cabin a drink,' he lied.

The pint pots came up fresh and clean from the galley and Jack, lurching to his feet to collect them, with great presence of mind, slipped the precious bottle back into his jacket pocket.

Leaning forward, he completely lost his balance, lunged for the door post, missed, caught his heel on the threshold of the storm sill, and crashed heavily backwards against the coal bunker in the corner. There was a sharp crack, a cry of pain and anguish and then an ominous dribbling sound.

Dick, quite quick to react, jumped to his feet.

'Now, you drunken old sod, 'ave you 'urt yerself?'

Jack, crumpled in the corner, coughed and gurgled. 'I feel I've got a sticky, wet arse – I hope it's blood!'

'Oh no, not the other bottle.' Dick was furious. 'I 'aven't 'ad a drink yet. Get up and let's 'ave a look,' dragging him to his feet with considerable strength born of desperation.

Jack reached into his jacket pocket and gingerly extracted the milk bottle by its neck. Miraculously, it was still almost three-quarters full but, as they peered at it short-sightedly, the whole of the lower half

which had been cracked right round the neck and hanging on by a hair, suddenly parted company and dropped, hitting the deck with a crash within inches of Parky's nose. He was left holding the top three inches of the neck, still tightly stoppered.

Parky recoiled in alarm and the two old gentlemen looked at one another in shocked disbelief. There was broken glass and gin swilling right across the deck of the tiny after wheelhouse.

Geoff, doing his best not to take any part in the pantomime that was unfolding, on hearing the second bottle hit the deck, poked his head round the corner.

'I say, chaps, is everything all right?' he breezed, and then, on seeing the chaos and mess on the floor, 'I'll get the lad along with a floor cloth and a bucket. We'll soon have the mess cleaned up.'

Seeing all his gin sloshing around was too much for Dick. 'Couldn't we try and scoop it up with summat, Jack?'

'Aye,' said Jack. 'Fetch us some spoons, Geoff, never mind the floor cloths, t'floor looks clean enough, we'll try and save some.'

'Watch out for glass, Mr Franks,' Geoff called as the two scrabbled and crawled about on hands and knees with spoons, scooping up the booze and ladling it into their pots.

'To hell with the glass, let's get the gin.' Old Dick chased the little rivulets as they moved from side to side with the gentle rolling of the cutter.

Old Parky, highly amused at their antics and not a bit pleased at being deprived his tipple, played up to them, offering words of encouragement.

'Try in the corner there, Mr Franks. Look, it's running this way, Mr Nixon.'

Jack, still rubbing his backside, babbled away, 'Get as much as yer can, never mind the glass, we can always strain it through summat later. OOOW! come over 'ere Dick – it's DEEEEPER!'

His excitement at rescuing something from the fiasco was overcoming any physical discomfort he might have felt.

Meanwhile, the cutter had arrived back alongside the jetty and Geoff, having supervised the mooring, shouted down to Parky, 'OK, Chief, that'll do,' and then, to the two crawling round in the after wheelhouse, he asked, 'I'm going to make a mug of cocoa, would anyone like a pot?'

Jack gave him a withering look but Dick's eyes twinkled and he nudged Jack in the ribs.

'That would be a very nice idea, Captain,' he responded. 'Mr Nixon and I could just drink a pot of cocoa. By the way, Captain,' continued Dick, all charm, 'You don't happen to have a clean hanky I could borrow?'

'Why, yes, I probably will have. I'll fetch it when I pop the kettle on.' Geoff disappeared down into the saloon and reappeared minutes later with a sparkling white, monogrammed handkerchief. Dick took the proffered item, muttered his thanks, and he and Jack continued their antics of sloshing and tinkling with various pieces of crockery. By now they had retrieved as much of the puddle as their spoons would allow and had almost a mugful of greyish brown liquid.

Speaking in lowered tones they sniggered and giggled at a private joke, all the time trickling and dribbling sounds being drowned by the occasional guffaw from Dick.

Geoff reappeared; two mugs of steaming, hot cocoa were thrust up through the hatch and placed on the threshold of the storm sill.

'Do you take sugar?' Geoff asked in his polite, precise way.

'Oh, don't worry, Captain,' Dick replied, his whiskery face beaming. 'We'll sweeten it to our taste up 'ere.' And grabbing the two pots, he whisked them round the corner into the aft compartment.

More mutterings and slurping sounds and after a while Jack shouted in a bawdy voice, 'By gum, Captain, this is a drop of good cocoa you've made us, I'll sleep like a baby after this.'

Clouds of cigar smoke filled the air as the two damp and crumpled old rogues finished their smokes with their drink and eventually heaved themselves to their feet, preparing to leave the cutter by way of the short iron ladder onto the jetty. Luckily it was almost high water so it was only a short climb. Eventually, with much cursing and muttering, the two hobbled up the jetty, occasionally colliding with each other as they staggered up side by side. Both arrived in the cabin doorway together almost jamming each other.

Bill Braithwaite, an enormous man with a florid complexion and a talent for plain speaking, looked up from his paper at the bedraggled pair.

'Where the hell have you two been? Oh, you've been drinking, I can smell it from here. Have you brought us a drop?'

'Well, actually,' Dick suddenly started speaking very posh. 'Well, actually, we did manage a little drink, but we spilt most of it. To be more precise, *he* spilt most of it,' indicating the swaying figure, all glassy-eyed, standing next to him.

Jack, with a great effort, tried to stand upright, his coat covered in filth from his grovellings. He suddenly clasped his hand to his forehead and moaned.

'Oooohh, I've come over all giddy. It must have been that hot cocoa Geoffrey gave us, Dick. I must have a little lie down until I feel better.'

And with that, he collapsed into the nearest iron cot, fully clothed, boots, raincoat and all, his greasy, old cap covering his face.

'Bluddy cocoa, indeed,' said Bill, and shook his paper in disgust.

CHAPTER 22

On the whole the *Lister*'s summer stay was looked upon by us as a bit of a vacation from the incessant open boat work we normally had to endure. Sometimes we would hear rumours and reports of how the work was progressing on the steam cutter up in the Dock. We would hear how a radio telephone direct link was being installed to the Harbour Master's Office. Our last remaining pulling boat had been fitted with a tiny two-stroke Stuart and Turner engine. Was technology finally catching up with us? We might even get a mechanical chipping hammer!

We looked upon the mechanisation of the old pulling boat with mixed feelings. Most things were better than rowing a heavy boat but, by all accounts, it was not fitted with an astern gear. However, it was only used when the weather was too bad to drop the bigger motor boat and the little Stuart engine was better than 'Armstrong's patent two-stroke', sometimes referred to as 'Norwegian Steam'. As to the RT link to the harbour office, this we didn't like as it meant keeping a radio listening watch in the wheelhouse and as such a direct curtailment of our freedom to roam the decks at night. We preferred the wink of the Aldis to the squawk of the radio. It would be all too easy to be at the beck and call of office wallahs.

We took the changes philosophically; after all, quite a few of us would soon be leaving to do our sea time, and then we would be out of it all. But here again the wiseacres on the Pilotage Authority had once more been at work. Instead of allowing us to approach tramp companies with the chance of an uncertificated third mate's job, as previous lads had done, they had arranged with certain Liner companies for us to sail with them as third and fourth year cadets.

This was a bitter pill for us after an impecunious four years in the boarding boats. Instead of looking forward to twenty-six pounds a month and all found, we would receive eight or ten pounds and still be the lowest form of 'marine life'. However, we didn't make the rules, and when we tendered a joint letter to the Authority voicing our side of the argument, we got a curt reply: 'If you don't like it you can always leave.' As they put it, we had been rough, horny-handed boat hands for long enough and they now intended we should go to sea as little gentlemen, have clean linen on the table and eat with a knife and fork! All protestations about the money side of it fell on deaf ears. Maybe the Board members had been financially secure too long to remember ever being thundering hard up. We grumbled and got on with the job.

The steam cutter arrived back on station, all prim and painted up like an ice-cream barrow. Of course some things didn't work as they should and had to be coaxed or bullied. The winch was a little stiff as the glands had all been re-packed and new bearings fitted. The newly caulked decks had to be scraped and scrubbed and whitened, our beautiful brass work was green and pitted after weeks of neglect. New lads had started and now we were the old hands who oiled the windlass and checked the battery acid level while the youngsters scrubbed floors and polished brasses; after all, at twenty we were nearly finished our boat hand days.

Once again from the cushy, passive role we had enjoyed on the *Lister* we were back in our boats, providing the vital link between floating base and incoming and outgoing shipping. We had to get used to the old pulling boat, now fitted with the little Stuart two-stroke. The engine had no gear-box, hence the lack of reversing power. It worked from a centrifugal clutch incorporated in a massive fly wheel. When starting and ticking over, the drive to the little two-bladed propeller was disconnected. On opening the throttle the fly wheel and the power transmitted directly to the propeller. Throttle back and the transmission line was broken again. All very simple but effective. The only thing left was for us to modify our techniques when going alongside ships. We had to keep giving bursts of throttle and make sure we didn't overshoot.

It was decided that as the opportunity offered, we would use the little boat as much as possible to get the hang of things before the winter clamped, and we had to use it in bad weather situations. The idea seemed all right except the boat proved painfully slow after our bigger boats and also, it had no weather protection, so it was confined to short trips in comparatively fine weather at first.

So it was one day in late October, the weather was fine and dry but a persistent north-easterly swell made the anchorage just inside the lighthouse rather uncomfortable. We rolled and rolled for several hours until the Pilot on turn had had enough so as there was nothing expected inward, we would move our position a mile or so up-river to a spot we all called Fifth Buoy hole. It was a good place to anchor at a point where the Seaton Snook channel met the main channel; from it we could see up river and also down to seaward. The Fifth Buoy light itself was quite an elaborate affair constituting a landing stage with steps leading up to a large, wooden deck area of several hundred square yards on which stood the light-keepers' living quarters and living accommodation. It was manned by two keepers who worked two days on and two days off constantly to keep the lights tended. These lights, a high light and a low one, were situated on the ends of very long rickety cat walks and were, in fact, the leading lights for vessels sailing up the main channel. To such a ship, the two lights would appear one above the other in a vertical line and, provided the vessel stayed on the mid-channel line, they would remain so until approached close to and the course had to be altered to pass them. They were an invaluable guide, especially to big ships, at night, and as such, in those days, warranted the two keepers to tend them at all times.

Once at anchor and snugly 'brought up' in a good spot with plenty of swinging space, the skipper sent us in the boat to the Fifth Buoy to use the 'phone and find out what sailings there were. We welcomed the chance for a chat with the old chaps and were invariably inveigled into a quick game of 'doms'. They took a great delight in beating us every time. Looking back, I suppose they must have had a great sense of tolerance towards each other, and as far as dominos were concerned, they had lived and worked together for so long, they were practically telepathic. Such was the state of their 'phone that we often had to wait half an hour or more before we could get through, and it gave us ample opportunity for mugs of tea and a bit of fruit cake.

Eventually we managed to find that the only ship sailing was a brand new tanker, *British Oak*, straight from the builders' yard and coming to sea to do engine trials, ballasting trials, and also adjust compasses and calibrate her DF.

I groaned – these jobs tended to go on and on and especially when compass adjusting; we had the Adjuster and his assistant to get out of the ship as well as the Pilot.

Some Adjusters – there were about three or four who operated in the area – were very little bother. But one in particular, Wes Billings, gave us no end of trouble, especially in blashy weather. This little man was a strange paradox of fearlessness and timidity. He was a swashbuckling private Pilot who flew his own 'plane with great panache, often buzzing us at zero feet with obvious glee, when we were anchored on station at the Fairway Buoy. He drove a big Chrysler Airflow with wild abandon, terrifying anyone who happened to hitch a ride with him up the long breakwater road. He talked loud and long about his various escapades and yet, once he cocked his leg over a ship's side and started descending the Pilot ladder, he was a complete fanny. It clearly terrified him and he often froze in sheer panic halfway down. Several times he had to be dragged bodily off the ladder, landing in the bottom of the boat in a heap, this being the only way to get him down into the boat if he funked it halfway, as no amount of shouting or cajoling made the slightest impression. Once dragged off the ladder, he would smooth his ruffled feathers, curse everyone and everything in sight and carry on as if nothing had happened.

We always made sure that the boat-hands' Christmas Funds contribution box was very obvious when we transferred him to the steam cutter. If he didn't get his hand down fairly generously, he got a wetting next time out!

All this and more was related to the fresh-faced lad sitting at the engine control as we motored back across the comparative calm of our haven anchorage. Young Derek had only been started a couple of weeks

Cutter Confusion

and was still a bit wide-eyed about everything.

On our way back to the anchored cutter, we passed a small flotilla of sand wherries, returning, deep laden with fine sand, to their base near the Transporter Bridge. These little, old, quaint craft always reminded me of the seven dwarfs; even their names had a dwarfish ring to them - *Sweep, Sophia, Hornet, Number 10, Rachel, Knockmaroon.*

When you really looked at them they were all different and yet when they arrived or departed from the edge of the sand where they beavered away with their clattering, hissing, steam grabs, they all had a look of sameness about them. Leaving their base just after high water, they came down on the ebb tide to arrive at the entrance to the Snook channel, worked over the low water period, and then, when their holds were full of sand, they would back off on the rising flood tide and steam gently back up to Middlesbrough again to discharge into lorries around high water time. I think at least two of them had originally been Clyde Puffers and the others had been adapted from long forgotten designs, some little more than powered barges.

No sooner were we back at the cutter than the new tanker was to be seen coming to sea in charge of her tugs. She looked a fine sight in the autumnal sunshine, her paintwork gleaming and clear cut, unsullied by scrub marks and rust streaks that would eventually mar her appearance. Her ensign and house flag fluttered brilliantly from her stern and mainmast head and above the sparkling white navigating bridge, latticed over with awning spars, flew the red and white 'H' flag and on a separate hoist the letters J I – 'I am manoeuvring to adjust compasses'.

As she approached, looking like a well groomed prima donna, she almost seemed as though trying to avoid breathing the black smoke that suddenly plumed from the for'ard tug's funnel. Someone from the bridge of the cutter hailed her as she passed a couple of hundred feet away. How long would they be adjusting compasses? The reply came that one turn each way around the Fairway Buoy should be sufficient. It was our old friend, the little man in a beret and a long mac – fearless Wes!

On the strength of that information we picked the boat up and prepared to go to sea. When the compass adjuster had finished he would want landing, but the Pilot had to stay a little longer while some engine adjustments were made and the ballasting trials completed. With luck we should be finished late in the afternoon. The head tug let go his tow line and swung away clear to port as the new ship set course past us, down the sea reach and out past the South Gare light. Our anchor slammed home in the hause pipe and with the boat bowsed in tight on the rail, we followed houndlike, in its mistress's footsteps.

The swell had subsided a lot with the coming of the flood tide and it was soon apparent that we should be able to anchor quite comfortably outside. This, in fact, was what we did, to lay handy and wait until the ship finished her pirouettes around the buoy. Then, at a signal from her siren and a hauling down of her signal hoist, we dropped the motor boat and ran to her. The ladder was not too long, ending about three feet short of the boat's gunnel, so it was an easy climb down for the Adjuster and his young assistant with their cases of heavy gear. Once in the boat and having recovered his composure, Wes told us the Pilot would be staying aboard while they pumped ballast and tested the engines in various states of trim. It could take three or four hours, then he would anchor and blow for us again. I grunted my acceptance of the situation and then cast a disapproving look at the lad at the engine as it began misfiring and coughing.

'Did you make sure there was enough juice?'

'Aye, the tank's nearly full.'

'Must be some muck in the petrol pipe, I'll have to blow it out when we get back.'

'How long will that take, son?' said Billings.

'Who can say,' says I, 'Depends how much muck is in the line. We've no compressed air to blow it through. Sometimes just blowing in the filler cap is enough to clear it, sometimes we have to take the whole lot adrift.'

'Hell, my missus is coming down to pick me up at one o'clock.'

'We'll be as quick as we can, Mr Billings. We could take you in the other boat but you might get very wet and it's painfully slow as well.'

'What about getting old rust bucket underway then?' He grinned at his calculated insult.

I grinned back. 'I think most of them have turned in for an afternoon's siesta, they won't budge 'til relief time.'

'Sod it then.' He spat the words into the wind and hunched his head into his shoulders.

The engine wheezed and coughed and faltered, picked up only to falter again. It had obviously swallowed something that hadn't agreed with it. We staggered back alongside and while the Adjuster fumed and fretted in the saloon, I went aft to find a few rusty tools with which to perform on the fuel line.

The engineer, whose job it really was, had turned in for the afternoon, so young Derek and I set to with stilston wrench, pliers and other instruments of the botch artist, to try and remove the union nuts on the fuel lines running under the engine bearers. It was obviously going to be a longer job than we anticipated so, as we hadn't eaten since breakfast, we decided to knock off to cook our lunch now that the galley was clear of Pilots.

This was just about the last straw for the impatient little man in the saloon, until I pointed out that maybe he'd eaten already, knowing how hospitable new Chief Stewards usually were when out to impress the office high brass. This seemed to quieten him a little and gave us time to eat our bubble and squeak and wash the taste of petrol out of our mouths with a pint of scalding tea.

I felt sorry for the young fellow who acted as the Adjuster's assistant, learning the business; he was the butt of his boss's jokes, his scapegoat if things weren't going quite right, his general dog's body and gopher. I never once heard him answer back or be the slightest bit disloyal to his master. When we asked him how he put up with being Wes's hand rag, he would just give a wry smile and say, 'Oh, the boss has his good points as well.'

The afternoon dragged on and, try as we may, we couldn't seem to get the engine to run smoothly. We tried the float chamber, the plugs and plug leads, fuel lines, everything we could think of within the scope of our limited mechanical knowledge. There is obviously a limit to what you can tackle with a few scruffy old tools in a boat jarping about in the water alongside the cutter. Finally we decided the only thing to do would be to pick the boat up out of the water and wind in the davits inboard and land it in its chocks on the boat deck. Here the engineer could work on it in a degree of comfort. Meanwhile we would have to run the job with the little motorised pulling boat. It was an unsatisfactory arrangement but a workable one. After changing over the clutches on the boat winch, we lifted the little boat and wound it out over the starboard side transferring the relieving tackles to the opposite side and moused all the tackle hooks to prevent them coming adrift when the falls and tackles were swinging free.

By the time all this had been done, the *British Oak* seemed to have completed her trials and had come to anchor about half a mile away. The Pilot blew a long blast and three short on the ship's whistle as a signal to us that he was ready to leave her, and we dropped the little boat in the water. The engine burst into life at the first swing of the handle with the characteristic pop, pop, pop of a two-stroke.

At the sound of the engine starting, the Adjuster appeared from the saloon alleyway.

'Are you running me ashore, lads?'

'No, Mr Billings, we're going to get the Pilot out of your ship. When we get back we'll get the cutter underway and go in for relief time.'

'And about time too,' Wes muttered under his breath.

Young Derek sat at the engine box and opened the throttle as we poppled away at about four knots across the half mile or so of water separating the two vessels at anchor.

It had been a fairly long day and I was looking forward to getting home about half past five and the rabbit stew that I knew was for tea. Rabbit was one of my favourites and the way mother cooked it with onions, carrots and mushrooms, and just a touch of curry powder to bring out the flavour, set my mouth watering just thinking about it. Then wash and change and a dash to catch the bus to meet a certain girl in town for second-house pictures. If the queue proved too long, maybe a walk along the stray or a cup of coffee in the High Street. All these thoughts filled my head as we approached the anchored ship.

The ballasting had drastically altered the *British Oak*'s trim. Instead of more or less even keel, as she had been three hours or so before, now she was right down deep in the water aft, and her clean painted bows were right out of the water for'ard. This trim was not a good one for us. As we swung round the stern and onto her lee side where the ladder hung like a plumb line down her tilted side, it was obvious that it was now about six or eight feet too short.

A pilot ladder – that much neglected essential to the Pilot's profession – holds a uniquely odd place amongst the deck equipment of a modern ship. In the normal course of events it is never ever used by any member of the crew, yet it is expected to be kept in top class condition, clean and sound and ready for use when needed.

It is nothing like the flimsy affair used by trapeze artistes and the like. It consists of inch-thick hard-wood rungs which are four and a half inches deep and about eighteen inches long. These rungs are spaced fourteen inches apart with a double thickness of two and a half inch manilla rope at each side passing through holes in the rungs or steps. To prevent the rungs losing their precise spacing vertically, they are secured above and below at each end by small, hard-wood, wedge-shaped chocks which are, in turn, seized to the double ropes in an altogether secure and solid manner.

Every ten feet or so is a much wider rung of about six feet, called a spreader. This is to prevent the ladder from turning round itself, especially with a Pilot on it. The whole design is to prevent, as much as possible, the ladder swinging about wildly, as, by its very nature, it is only made fast at the top to the ship's bulwark rail.

The only way of handling it or storing it, when not in use, and this can be for long periods between ports, is to roll it up like a stair carpet.

In those days, the practice was to have a reasonably long ladder of, say, eighteen to twenty feet and, if the vessel's free board demanded it, a shorter piece of say twelve feet, which could be shackled onto the longer one as required.

As we approached the ship's side, the Pilot appeared on the wing of the bridge and called out that he would be a couple of minutes.

'It's no good, Mr Swinbanks,' I yelled back. 'The ladder's miles too short now. You'll have to shackle another length on.'

A pea whistle blew on the ship's bridge and men scampered along the deck to answer it. Eventually three or four riggers from the shipyard appeared and began unlashing the ladder from the rail, then two of them reached over and began hauling it, slowly at first, back up the ship's gleaming side.

Derek and I put-puttered around in the boat to wait while the riggers on deck manhandled a stiff, new ladder from the mast house and along the deck. We waited three or four feet from the ship's side. Young Derek was very keen on racing pigeons and I was pulling his leg about the various merits of rabbit pie and pigeon pie. He laughed and his face flushed a little at the jibes as he sat on top of the engine box facing me.

I heard a cry from above. The laugh on my companion's face froze. His eyes flickered with sudden alarm. A great snake of heavy ladder whistled as it unrolled in mid-air and scythed out and down like the flick of a whiplash.

Instantaneous is a word hard to relate to.

I never knew what hit me . . .

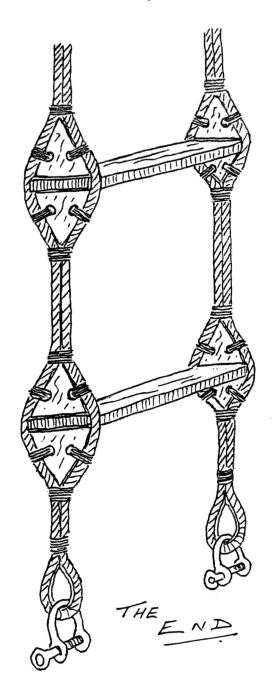

5